Mechanical Engineering Series

Frederick F. Ling
Series Editor

Springer
New York
Berlin
Heidelberg
Barcelona
Budapest
Hong Kong
London
Milan
Paris
Santa Clara
Singapore
Tokyo

Mechanical Engineering Series

Introductory Attitude Dynamics
F.P. Rimrott

Balancing of High-Speed Machinery
M.S. Darlow

Theory of Wire Rope
G.A. Costello

Theory of Vibration: An Introduction, 2nd ed.
A.A. Shabana

Theory of Vibration: Discrete and Continuous Systems, 2nd ed.
A.A. Shabana

Laser Machining: Theory and Practice
G. Chryssolouris

Underconstrained Structural Systems
E.N. Kuznetsov

Principles of Heat Transfer in Porous Media, 2nd ed.
M. Kaviany

Mechatronics: Electromechanics and Contromechanics
D.K. Miu

Structural Analysis of Printed Circuit Board Systems
P.A. Engel

Kinematic and Dynamic Simulation of Multibody Systems: The Real-Time Challenge
J. García de Jalón and E. Bayo

High Sensitivity Moiré: Experimental Analysis for Mechanics and Materials
D. Post, B. Han, and P. Ifju

Principles of Convective Heat Transfer
M. Kaviany

(continued after index)

Krishna C. Gupta

Mechanics and Control of Robots

With 38 Illustrations

Krishna C. Gupta
Department of Mechanical Engineering (M/C 251)
University of Illinois at Chicago
842 West Taylor Street
Chicago, IL 60607-7022 USA

Series Editor
Frederick F. Ling
Ernest F. Gloyna Regents Chair in Engineering
Department of Mechanical Engineering
The University of Texas at Austin
Austin, TX 78712-1063 USA
and
William Howard Hart Professor Emeritus
Department of Mechanical Engineering,
Aeronautical Engineering and Mechanics
Rensselaer Polytechnic Institute
Troy, NY 12180-3590 USA

Library of Congress Cataloging-in-Publication Data
Gupta, Krishna C.
Mechanics and control of robots/Krishna C. Gupta.
p. cm. — (Mechanical engineering series)
Includes bibliographical references and index.
ISBN 0-387-94923-2 (hardcover:alk. paper)
1. Robotics. 2. Robots—Control systems. 3. Automatic machinery.
I. Title. II. Series: Mechanical engineering series (Berlin, Germany)
TJ211.G86 1997
629.8′92–dc21 96-37619

Printed on acid-free paper.

Production managed by Hal Henglein; manufacturing supervised by Johanna Tschebull.
Camera-ready copy prepared from a LaTeX file.
Printed and bound by Maple-Vail Book Manufacturing Group, York, PA.
Printed in the United States of America.

9 8 7 6 5 4 3 2 1

ISBN 0-387-94923-2 Springer-Verlag New York Berlin Heidelberg SPIN 10560905

Series Preface

Mechanical engineering, an engineering discipline born of the needs of the industrial revolution, is once again asked to do its substantial share in the call for industrial renewal. The general call is urgent as we face profound issues of productivity and competitiveness that require engineering solutions, among others. The Mechanical Engineering Series features graduate texts and research monographs intended to address the need for information in contemporary areas of mechanical engineering.

The series is conceived as a comprehensive one that covers a broad range of concentrations important to mechanical engineering graduate education and research. We are fortunate to have a distinguished roster of consulting editors on the advisory board, each an expert in one of the areas of concentration. The names of the consulting editors are listed on the following page of this volume. The areas of concentration are applied mechanics, biomechanics, computational mechanics, dynamic systems and control, energetics, mechanics of materials, processing, production systems, thermal science, and tribology.

Professor Marshek, the consulting editor for dynamic systems and control, and I are pleased to present this volume of the series: *Mechanics and Control of Robots*, by Professor Gupta. The selection of this volume underscores again the interest of the Mechanical Engineering Series to provide our readers with topical monographs as well as graduate texts.

Austin, Texas   Frederick F. Ling

Mechanical Engineering Series

Frederick F. Ling
Series Editor

Preface

This book has evolved from a graduate level course in robotics that the author has taught for over a dozen years at the University of Illinois at Chicago. It is addressed primarily to first-year graduate students in mechanical engineering. However, over time, this course has also attracted students with a wide variety of academic backgrounds, e.g., mechanical, industrial, electrical, and bio-engineering. For many of these students, this is even the first exposure to topics in spatial kinematics and dynamics. Therefore, the content of the book has been kept at a fairly practical level. Many advanced concepts are briefly explained and their use emphasized, but the related theory and complicated formal proofs have been omitted. It is hoped that after studying some important concepts in kinematics, dynamics, and control in the context of robotics, students will be motivated to study these specialized subjects further in their graduate studies. Topics that have been selected are of contemporary interest in the field. An attempt has been made to expose the students to a broad range of topics and approaches. Some of the material presented is based upon the author's own research in the field since the late 1970s.

Chapter 1 contains general preliminaries. It also emphasizes the distinction between the passive (or coordinate transformation based) approach and the active (or spatial displacement based) approach. In the former, the world is viewed as a series of coordinate transformations applied to a typical point, while in the latter it is viewed as a sequence of physical rigid body displacements, all occurring in the base coordinate system. The two approaches are related, but it turns out that the related mathematical details are quite different. This book is devoted to presenting the active displacement approach as a comprehensive approach to robotics, and that may be a unique feature of this introductory level graduate textbook. Both the passive and active approaches are presented in sufficient detail so that the students can learn to work simultaneously with both approaches. Although the active displacement approach is based upon a more complicated theory of general spatial displacements, it is surprising that the end results are quite simple to apply, and students with even minimal prior exposure to spatial motions learn to do this rather quickly. A compromise had to be made, however, in the level of presentation, which emphasizes primarily the how-to-do-it aspect and not the very formal aspects of the theory.

Chapter 2 presents the widely used Pieper–Roth method (1969) and the zero-reference-position method (1981) for inverse kinematic analysis. The former method is based upon the passive approach and utilizes the pioneering contribution of Denavit–Hartenberg (1955) for representing spatial linkage systems. The idea of decoupling, which occurs in the inverse position solution due to the presence of a spherical wrist, is also emphasized; in a spherical wrist, the revolute joints associated with the wrist motion have cointersecting axes. The latter method, designated as the ZRP method, is based upon the active displacement approach and has evolved as a comprehensive alternative method for robotics. Although the ideas for using the active displacement approach for doing robotics calculations have been floated in the literature from time to time, this book attempts to formulate and present these in a single source that can be used as a textbook or a reference book for an introductory-level graduate course in robotics. This chapter also formulates the differential kinematics and iterative position analysis based upon the conventional Newton–Raphson based approach and the unconventional ODE-based approach.

Chapter 3 is devoted to workspace characterization and determination. The problem of determination of primary or dexterous workspace for general robots remains an open problem, but special cases are presented where this can be done. An important result related to tool-spin capability is also presented.

Consequences of decoupling due to a spherical wrist are discussed in Chapters 2 and 3. Spherical wrist construction and its extensions into a variety of novel bevel geared wrists have emerged as important features in modern robot manipulator designs.

Chapter 4 presents an extremely brief discussion of dynamics and control. An attempt is made to review the basic Newton–Euler and Lagrange approaches and demonstrate how these can be adopted for the active displacement framework utilized for robot kinematics in the earlier chapters. The section on control presents the computed-torque method in which a combination of feedforward and feedback signals is utilized to render the system error dynamics into a set of decoupled second-order linear ODEs.

Readers should also review a special explanation in Section 4.4 on mixing of vector and matrix operations. This mixing has been done to avoid using the formalism of tensors or dyadics.

An extensive bibliography has been provided so that interested readers can pursue the literature on many topics presented in the book. It should also provide enough material to support several term papers and projects by graduate students. Readers can also find newer material by using the papers included in the References as search sources for *Science Citation Index.*

The author would like to acknowledge the collaborations with his friend, professional colleague, and former professor, Bernard Roth, and with the author's former students, Jim Hansen, Gary Carlson, Kazem Kazerounian, Salem Samak, Clifford Mirman, Hui Cheng, Xiaoming Chen, Vinod Singh, and Rufei Ma. Also acknowledged is the professional assistance by the staff of Springer-Verlag, New York, especially by Dr. Thomas Von Foerster. Finally the author appreciates the patience and sacrifice by his wife, Karuna, and daughter, Anupama, over the many years during which this book was developed and written.

K.C. Gupta
Chicago, Illinois
December, 1996

Contents

1
Introduction

1.1 Background

Robots as humanoids have been depicted in science-fiction literature and movies for many decades. These humanoids tended to have some human characteristics such as physical dexterity, speech, intelligence, and emotions. The industrial robots, as we know them today, do not resemble such humanoids. In fact, they look more like a complex system of machinery.

The pioneering work in robotics was done in the 1960s in the academic laboratories at MIT and Stanford. Early commercial robots became available in the mid-1970s. These early industrial robots had simple mechanical structures that attempted to actuate the familiar coordinate systems such as Cartesian, cylindrical, and spherical; had combinations of electric, hydraulic, and pneumatic drives; and had primitive teach-and-play type controllers. Subsequent industrial robots had more complex arm-wrist structures, had all-electric drives, and had improved programming capabilities. Several companies in the United States, Japan, and Europe entered the robotics business, and many articles in the popular press were written about the coming robot revolution, fully automated factories, and a new philosophy of manufacturing. The economic downturn in the early

1980s hurt the robotics industry very deeply, and, at least in the United States, it did not fully recover from that shock. By the late 1980s, the U.S. robotics industry was in total disarray, and all major robot manufacturing operations in the U.S. were either dissolved or taken over by foreign companies. The major players in the commercial robot markets today are Japanese and European companies. Yet, there continues to be a great interest in research and development related to robots in U.S. academic institutions, much more so than even in Japan and Europe.

Conventional mass production is based upon fixed automation assembly lines. Large quantities of good quality products can be produced at affordable prices by mass production. A minor product changeover is made possible by temporarily shutting down the assembly line and adjusting some of the tooling and assembly line flows. However, a major product changeover requires costly retooling and complete rebuilding of the assembly line. The customer has limited choices of product selection and options under business monopolies, but the competition ensures a reasonable range of product selection and options—the time when Henry Ford could say that the customer could choose any color, so long as it was black, is long past.

Industrial robots are, to varying degrees, programmable systems, and they make flexible automation possible. In a flexible manufacturing system, it is possible to reprogram the robots and the product flows, and the same flexible assembly line can produce a wide range of products. A family of products with a wide range of options can also be produced on demand. New concepts such as economical small-batch production, just-in-time manufacturing, and on-demand production have not only revolutionized manufacturing but how the products are marketed and sold. Instead of mass production, the robot revolution makes mass customization possible. For example, a Japanese bicycle manufacturer can produce over eleven million variations to personalize its bicycles to suit its customers' physiques and tastes; a major U.S. electronics manufacturer, based in Chicago, can produce twenty-nine million variations in its line of pagers to suit its customers' needs; a major jeans producer/retailer is experimenting with personalized measurements and order taking, electronically transmitting the order to a centralized robotic production line, and then shipping the order directly to the customer in a few days. One could not think of such production flexibilities and wide-open customer choices just a decade ago.

A fully automated factory, which requires minimal human supervision and intervention during a normal shift, has become possible. It is also intriguing to think about fully automated robotic factories producing new and improved robots—robots making more robots.

Industrial robots can be used in applications that require repetitive tasks over long periods of time, operations in hazardous environments (nuclear radiation, under water, space exploration), and precision work with a high degree of reliability. They can also be used by handicapped persons to overcome some of their physical disabilities.

There are some disadvantages that must be considered whenever a robotic application is contemplated. The initial costs of installing robotic systems are high. The operators, who must be technologically sophisticated, require special training. The maintenance and repair are difficult as well as costly. Studies have also shown that in a majority of applications where robots have been employed, the robots continue to do the tasks that they were initially programmed to do; the need to reprogram the robots did not materialize in these applications. A careful assessment of production needs is necessary to avoid such misapplications of robots because, in the instances mentioned, a fixed automation solution would have been more economical. Robots in factories can pose a danger to human operators and workers, even to other machines and robots, and this must be considered in the planning and design of the flexible manufacturing system.

1.2 Elements of a Robotic System

The main elements of a robotic system are manipulator arm-wrist, sensors, control system, and computer. The sensing can include joint position and velocity sensing, force sensing and/or touch/proximity sensing at the wrist-gripper, and vision. The control may range from low-level servo control to higher level model-based intelligent control. The computer may be a dedicated microprocessor-based hardware, a personal computer, or a large central computer that is also carrying on other computing functions.

The links of a robot are connected by means of kinematic joints

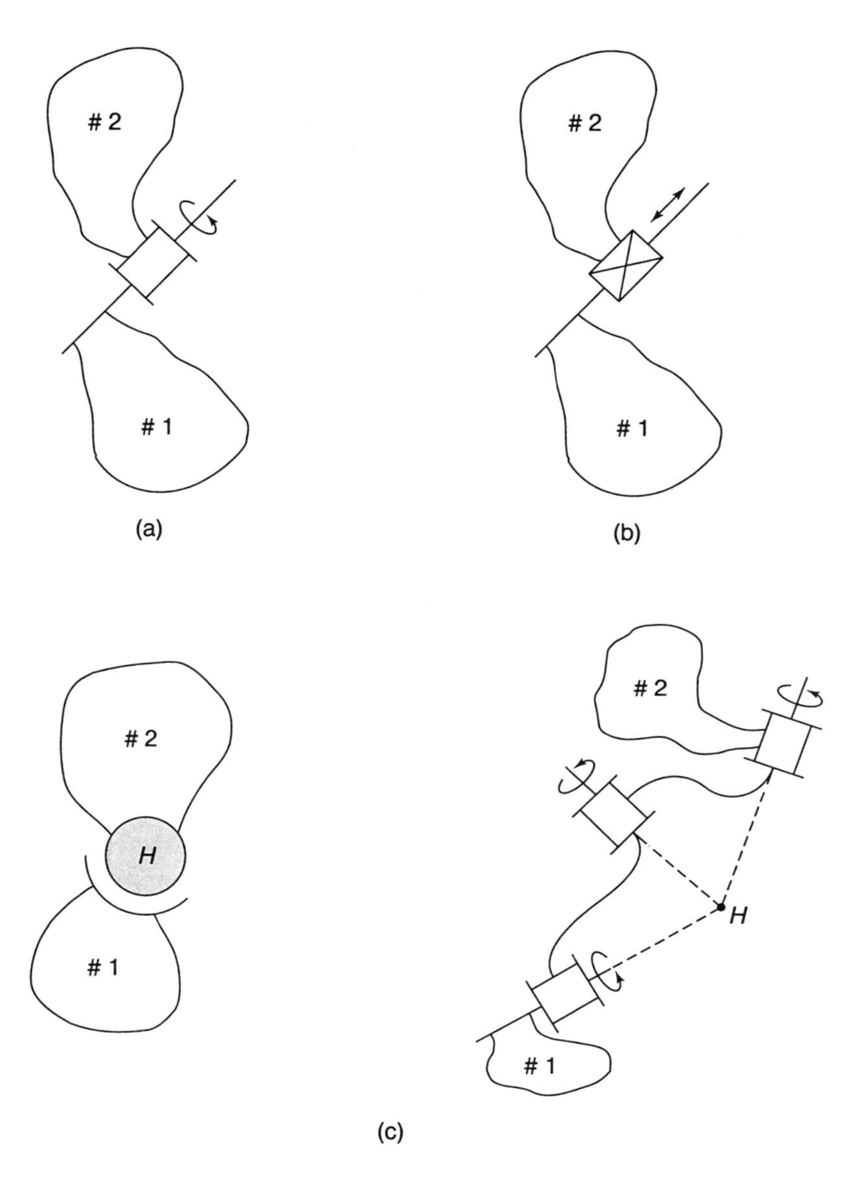

FIGURE 1.1. (a) Kinematic representation of a revolute (R) joint. Not shown is the simple axial view representation as a circle with a pin at the center. (b) Kinematic representation of a prismatic (P) joint. Not shown are common representations such as a rod in a tube, etc. (c) Kinematic representation of a spherical joint and its equivalent three-revolute construction.

that permit relative motion (Figure 1.1). The two common types of joints are revolute (R) and prismatic (P). The kinematic representation of a revolute joint [Figure 1.1(a)] is a rectangular element with two end-stop lines to denote that only rotational motion with respect to the central shaft element is possible. The end-stop lines indicate that sliding is excluded. The kinematic representation for a prismatic joint [Figure 1.1(b)] has a rectangular element with two diagonal lines that exclude rotation, and only sliding is possible (no end-stop lines are shown). The kinematic representation for a spherical joint [Figure 1.1(c)] is by a ball-and-socket joint. In the sketch of a kinematic chain, care must be exercised by showing the link connections across the joints and that proper relative motions should be possible at the joints; see Figure 1.1(c), for example, where three serially connected revolute joints are between bodies 1 and 2. Let us define the degrees-of-freedom (dof) of the space as d and that of the ith joint as f_i, and the constraints associated with the ith joint as u_i. Then, for planar motion (loosely called two-dimensional motion), $d = 3$; for spherical motion (three-dimensional motion with a fixed point), $d = 3$; and for general three-dimensional motion, $d = 6$. For each joint, the following relation must hold: $d = u_i + f_i$. The revolute joint permits relative rotation with one degree-of-freedom ($f = 1$, $u = 5$), and it can be thought of as a three-dimensional "hinge." The prismatic joint permits relative translation with one degree-of-freedom ($f = 1$, $u = 5$), and it can be thought of as a "square shaft in a square hole." The revolute and prismatic joints can be actuated by means of 1 dof pneumatic, hydraulic, or electric motors. A system of gears or rack-and-pinion or tendons can be a part of the drive system. There are other types of kinematic joints which have more than one degree-of-freedom, such as C—cylindrical ($f = 2$, $u = 4$), and S—spherical ($f = 3$, $u = 3$), but these are difficult to actuate.

However, the spherical joint is conceptually significant. A spherical joint can be thought of as a ball-and-socket joint [Figure 1.1(c)], e.g., our shoulder and hip joints. Let the center of the spherical joint be point H. Then, its three rotational degrees-of-freedom can be modeled by three revolute joints in series whose axes cointersect at point H. In this model, rotations about the three revolute joint axes do not affect the position of point H, and the basic condition for the spherical joint at joint center H is satisfied. If the spherical joint is unrestricted in its range of three rotational motions, which is not possible physically, then its theoretically equivalent revolute model is cointersecting $R \perp R \perp R$ (see Section 3.6). Notation $\perp$ between two joint symbols indicates that their joint axes are perpendicular, and notation $\|$ will indicate that two joint axes are parallel. Many robots have wrists whose joints can be modeled as an equivalent spherical joint; point H is then called the spherical wrist center. The human wrist can also be approximately modeled as a spherical joint.

1.3 Degrees-of-Freedom of a System

Let the total number of links, including the base, be n and the number of joints be g. Then, the degrees-of-freedom (dof) F of a mechanical system are given by the Grubler–Kutzbach criterion

$$F = d(n-1) - \sum_{i=1}^{g} u_i. \tag{1.1}$$

This equation gives the number of independent actuators that are needed to drive the system. It is desirable that these system degrees-of-freedom result in those of object manipulation by the robot end-effector. This may or may not be the case due to specialized robot construction or redundancy in actuation. For general purpose manipulators, the requirement is that $F \geq d$. Another difficulty that may arise during the movement of an object by the robot is that the object may reach locations where its translations in certain directions, or rotations about some directions, may not be possible. The degrees-of-freedom for object manipulation are then less than d even for general purpose robots, and such positions of the robot are identified as singularity positions. Singularities will be discussed in more detail in Chapter 2.

Replacing $u_i = d - f_i$, an alternate form is obtained:

$$F = d(n-g-1) + \sum_{i=1}^{g} f_i. \tag{1.2}$$

In a serial manipulator [Figure 1.2(a)], the links are connected in a chain-like serial fashion, and $(n-g-1) = 0$. Thus, a serial manipulator with six single degree-of-freedom joints will, in general, have six degrees-of-freedom for the gripper motion; exceptions may

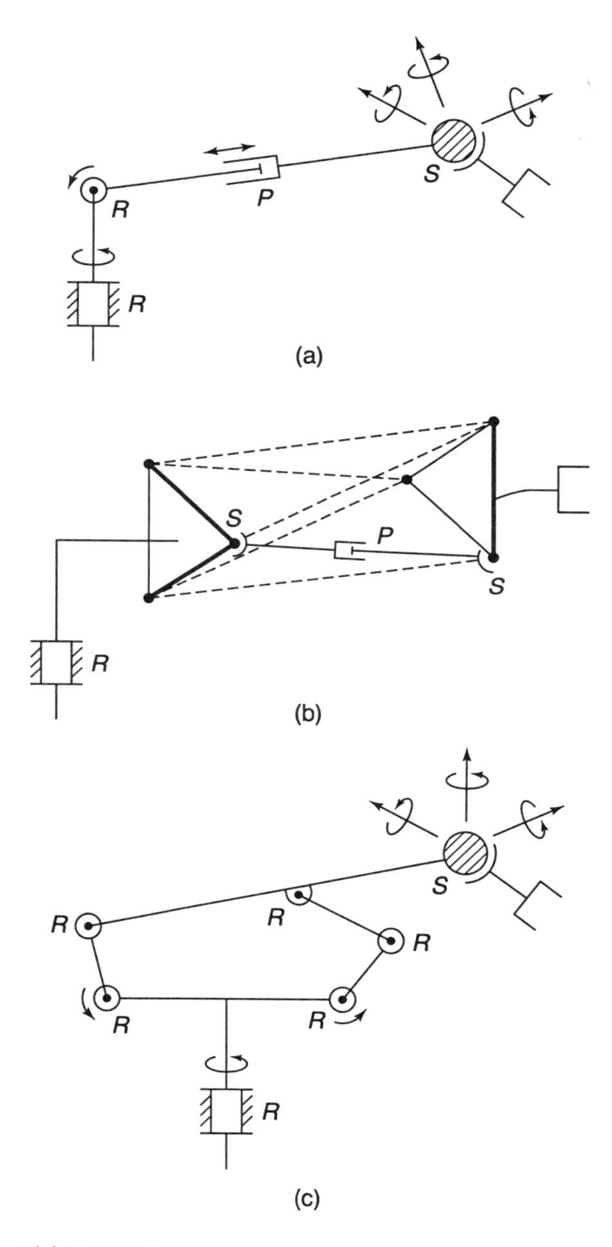

FIGURE 1.2. (a) A serial manipulator with joints occurring sequentially. (b) A parallel manipulator with a typical linear actuator shown; other linear actuators are indicated by dashed lines. (c) A hybrid manipulator with a closed substructure.

arise if, for example, more than three prismatic joints are used, three prismatic joints are coplanar, or more than three adjacent revolute joints have cointersection property (or are parallel—the cointersection point then is at infinity). Serial manipulators have large work zones (workspaces) for their grippers but suffer from the problem of having large masses that are far from the base column. At high speeds, they can have poor dynamics, significant vibrations, and problems due to elastic deformations.

In parallel manipulators [Figure 1.2(b)], a moving platform is connected to a fixed base by means of several linear actuators that "act in parallel," or "act in concert," or act simultaneously; the word parallel is not used here in its geometric sense. The linear actuators are connected to the base and the moving platform either by means of S-spherical joints or intersecting $R \perp R$ pairs. Parallel manipulator structures have rather limited work zones (workspaces) for the gripper, but they are quite sturdy. Two related applications are the airplane flight simulators and the vehicle dynamics simulators. The degrees-of-freedom can be found from Eq. (1.1), but an alternate method is as follows. If n' is the number of parallel connections (sometimes called legs), and f' is the net number of degrees-of-freedom between the moving platform and the base that is provided by a "leg," then the degrees-of-freedom of the moving platform can also be found as follows [in Eq. (1.2), set $n = 2$, $g = n'$ and $f = f'$]:

$$F = \sum_{i=1}^{n'} f'_i - d(n' - 1). \tag{1.3}$$

In addition, any two "legs" form a closed kinematic loop, and each such loop should have a loop degree-of-freedom that exceeds the overall degree-of-freedom F,

$$f'_i + f'_j - d \geq F, \quad i \neq j. \tag{1.4}$$

Some manipulators have a combination of serial and parallel structures, and these are called hybrid structures [Figure 1.2(c)]. Caution must be exercised in applying these equations to systems that contain closed substructures with special kinematic properties such as parallel or cointersecting axes because then $d = 3$, instead of $d = 6$, must be used for these substructures.

1.4 Regional and Orientational Structures

In six degrees-of-freedom serial robots, it is convenient to think of a regional structure [Figure 1.3(a)] consisting of the first three joints (inboard joints) and an orientational structure [Figure 1.3(b) and (c)] consisting of the last three joints (outboard joints). In robot arms with spherical wrists, the regional structure influences the positioning of the wrist center H (i.e., gross robot motion), and the orientational structure can place the gripper in a desired orientation about the wrist center H (i.e., fine robot motion). There are eight possible structures for the three degrees-of-freedom regional structure: PPP, PPR, PRP, RPP, PRR, RPR, RRP, and RRR.

A Cartesian regional structure has $P \perp P \perp P$ placed along the three XYZ coordinate directions. A cylindrical regional structure models the θzr cylindrical system by using $R \parallel P \perp P$. A spherical regional structure models the $\theta \phi r$ spherical coordinate system by using cointersecting $R \perp R \perp P$. None of these three structures can provide a capability to bend around obstacles, which articulated structures with three revolute joints can provide. A simple articulated arrangement is patterned after the human arm and has $R \perp R \parallel R$ such that the first two revolutes also intersect and form a 2-dof "shoulder" (note that the human shoulder really has three dof), while the third revolute forms a 1-dof "elbow."

Some early wrists did not have the cointersection property [Figure 1.3(b)] and included a small amount of offset between the last two wrist axes (e.g., Cincinnati Milacron T^3 robot wrist), but there are several advantages to having wrists that can be modeled as spherical joints [Figure 1.3(c)]. These advantages will be mentioned when kinematic analyses and workspace dexterity are discussed. Wrists with cointersecting $R \perp R \perp R$ will have mechanical interference when large rotations are made about the middle joint. It is often desirable to have wrists with a continuous three-roll property, i.e., the requirement that all three wrist joints can turn 360° without any mechanical interference. One way to achieve this is to retain the cointersection property but tilt the middle wrist axis so that the wrist angles become $90° + \beta$ and $90° - \beta$ [second arrangement in Figure 1.3(c)]. Another way is to keep the angles 90° but to give up the cointersection property by introducing a small amount of offset along the middle joint [third arrangement in Figure 1.3(b)].

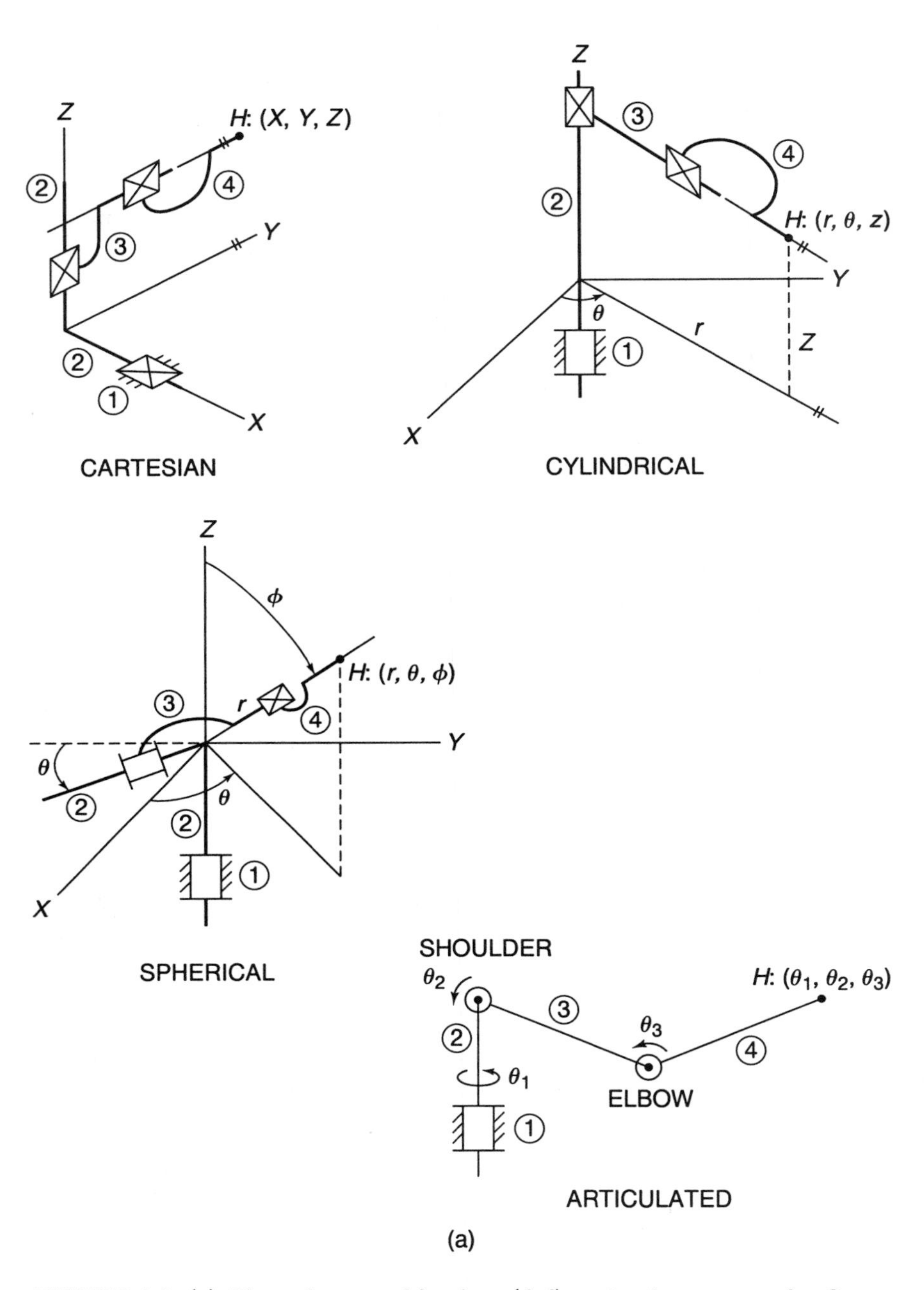

FIGURE 1.3. (a) Three degrees-of-freedom (dof) regional structures for Cartesian, cylindric, spherical, and articulated positioning of point H.

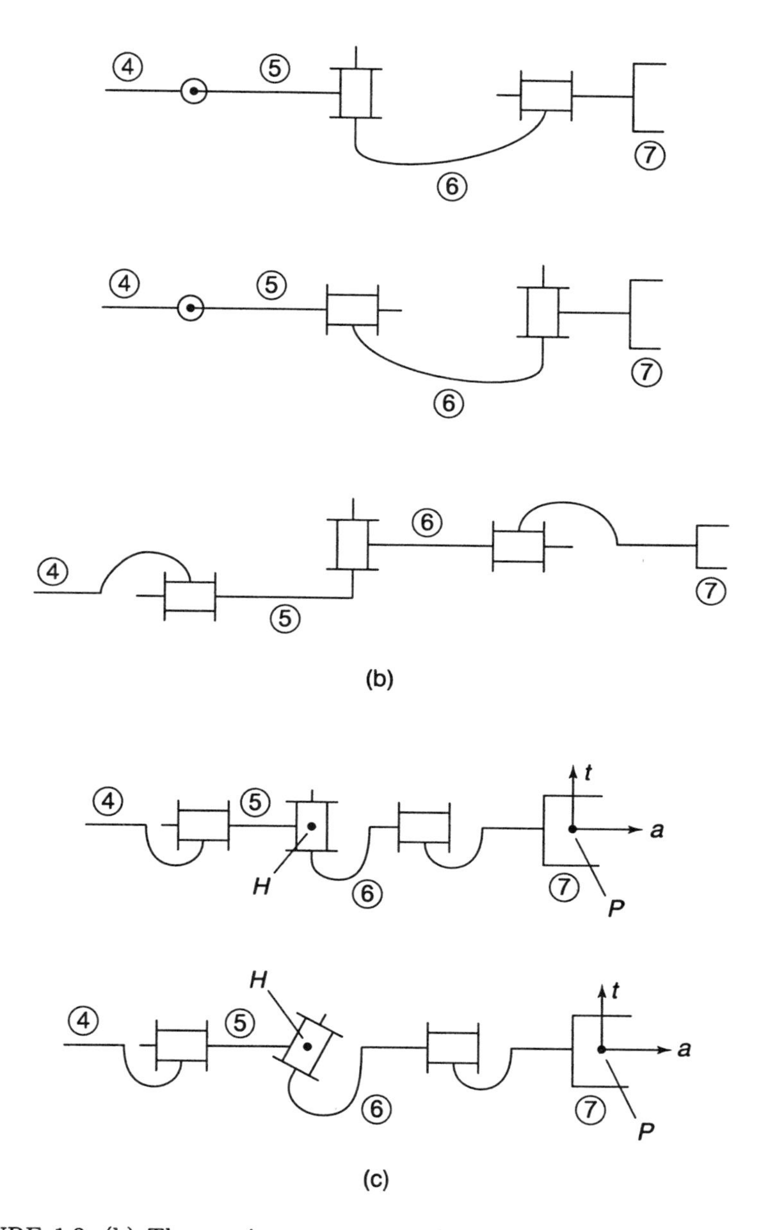

FIGURE 1.3. (b) Three wrist structures with nearly cointersecting axes. The last wrist structure can provide a continuous (or complete) three-roll. (c) Two wrist structures with cointersecting axes, i.e., "spherical" wrists. The last wrist structure can provide a continuous (or complete) three-roll.

1.5 Coordinate Systems

Robotic systems are complex systems, which require many coordinate systems to describe their physical structure and the task that is to be performed. On a factory floor, there may be other machines and robots, and their interactions can best be described in a "world" coordinate system, which can be fixed at a convenient spot on the factory floor. The action of each robot can be described meaningfully in a "base" coordinate system, which is attached to the base column of the robot. The relative motion between the links of a robot can be described in terms of the "link" coordinate systems, which are attached to various links; these link systems may play a crucial role in the kinematic dynamic analyses and control of the robotic system. The task that the robot is to perform can be described in terms of the "hand" coordinate system or the "tool" coordinate system if the hand (or gripper) is holding a tool to perform the desired task. A clear understanding of the various coordinate systems, and transformations among them, is necessary for carrying out analytical work in robotics.

Active and Passive Approaches

Let us consider two related but different situations. For simplicity of discussion, we will look at the two-dimensional case. First, we consider the fixed base coordinate system XOY and an object that is in motion. The moving object is being viewed from the fixed base system. This is called the active approach and it deals with physical displacements. At time t_1, a body point in position 1 of the moving body has base coordinates (X_1, Y_1), and at time t_2, the same body point in position 2 of the moving body has base coordinates (X_2, Y_2); both coordinates are measured in the base coordinate system XOY and essentially define position vectors $\mathbf{P}_1$ and $\mathbf{P}_2$ in the base system. Let the body rotation between times t_1 and t_2 be ϕ and the displacement of a special body point that is coincident with the base system origin at time t_1 be (d_x, d_y). Then, we have the following relation between the two sets of base coordinates (or the position vectors):

$$\begin{bmatrix} X_2 \\ Y_2 \end{bmatrix} = \begin{bmatrix} d_x \\ d_y \end{bmatrix} + \begin{bmatrix} \cos\phi & -\sin\phi \\ +\sin\phi & \cos\phi \end{bmatrix} \begin{bmatrix} X_1 \\ Y_1 \end{bmatrix}. \tag{1.5}$$

This can also be written in "homogeneous" form as

$$\begin{bmatrix} X_2 \\ Y_2 \\ 1 \end{bmatrix} = \begin{bmatrix} \cos\phi & -\sin\phi & d_x \\ \sin\phi & \cos\phi & d_y \\ 0 & 0 & 1 \end{bmatrix} \begin{bmatrix} X_1 \\ Y_1 \\ 1 \end{bmatrix} \tag{1.6}$$

or

$$\begin{bmatrix} X_2 \\ Y_2 \\ 1 \end{bmatrix} = [D_{1\to 2}] \begin{bmatrix} X_1 \\ Y_1 \\ 1 \end{bmatrix}. \tag{1.7}$$

The matrix $[D_{1\to 2}]$ is called the displacement matrix; it displaces the point from position 1 to position 2 (or changes position vector $\mathbf{P}_1$ into $\mathbf{P}_2$). The only coordinate system used is the base XOY system. Note that a special point $P_1\colon X_1 = Y_1 = 0$ is displaced to $P_2\colon X_2 = d_x$, $Y_2 = d_x$.

The homogeneous form includes both the "rotational" and "translational" information in a single matrix, which is the displacement matrix. Note that the extra equation is really $1 = 1$. This representation is advantageous for combining several displacements into a resultant displacement.

Second, we consider the situation when there are two coordinate systems, XOY and xoy, such that the angle between the X and x axes is ϕ, and the coordinates of the origin o in the XOY system are (X_o, Y_o). Then, a particular point P has coordinates (X, Y) in the XOY system and (x, y) in the xoy system. The same physical point P is being viewed from two different coordinate systems. This is called the passive approach, and it deals with coordinate transformation. The relation between the two sets of coordinates of P is

$$\begin{bmatrix} X \\ Y \end{bmatrix} = \begin{bmatrix} X_0 \\ Y_0 \end{bmatrix} + \begin{bmatrix} \cos\phi & -\sin\phi \\ +\sin\phi & \cos\phi \end{bmatrix} \begin{bmatrix} x \\ y \end{bmatrix}. \tag{1.8}$$

These relations follow immediately from the X and Y projections of the x and y coordinates of point P. These can also be written in homogeneous form as

$$\begin{bmatrix} X \\ Y \\ 1 \end{bmatrix} = \begin{bmatrix} \cos\phi & -\sin\phi & X_0 \\ \sin\phi & \cos\phi & Y_0 \\ 0 & 0 & 1 \end{bmatrix} \begin{bmatrix} x \\ y \\ 1 \end{bmatrix}. \tag{1.9}$$

Note that the third column in Eq. (1.9) has the X and Y coordinates of the origin "o" of the xoy system; this is unlike the special displacement vector $\mathbf{d}$ in Eq. (1.6). These two relationships in Eqs. (1.6) and

(1.9) are quite similar, and there should be some connection between these two approaches (active and passive). The last matrix columns in Eqs. (1.6) and (1.9) have different numerical values and physical meanings. We will consider the active approach to be fundamental and deduce the passive approach from it.

In the first case (i.e., active case), attach the coordinate system xoy to the moving body so that its current position (CP) is described by $(X_o, Y_0; \phi)$. Also consider a base coincident position (BCP) of the moving body when xoy and XOY systems coincide with each other. Then, the coordinate transformation matrix in the second (passive) approach is the special displacement matrix $[D_{\text{BCP}\rightarrow\text{CP}}]$. That is, the coordinate transformation matrix that changes the coordinates from the xoy system to the XOY system is also the special displacement matrix that moves a body from the base coincident position (BCP) to the current (CP) position of system xoy. Only in this special case, the last matrix columns in Eqs. (1.6) and (1.9) are the same; otherwise, the two columns are different. In three dimensions, the sizes of the matrices change, but the fact that the coordinate transformation is a special kind of displacement remains valid.

Denavit and Hartenberg (DH) Systems and Matrices

If link coordinate systems are chosen arbitrarily, then it will take six dependent variables to describe the relationship between two adjacent link coordinate systems. Constraints representing joint connections must be imposed explicitly. A widely used minimal system to represent links connected with axial joints (R, P) is the Denavit and Hartenberg (DH) coordinate system. Unfortunately, many minor variations of the original DH system ("DH-classic") can be found in textbooks and papers; besides causing confusion, these variations do not really contribute anything new to the topic. Therefore, it is very dangerous to use the final equations from a book or paper without first determining which variant of the DH system has been employed and whether the user is working with an identical DH system. We will describe the "DH-classic" system here.

Links are numbered in an ascending order from the base link, which is link 1. Link i [Figure 1.4(a)] carries joint axes $(i-1)$ and i; alternately, link i is the physical body that exists between the joints $(i-1)$ and i. The actual shape of the link is not important for kinematic analyses as long as the following two dimensional parameters

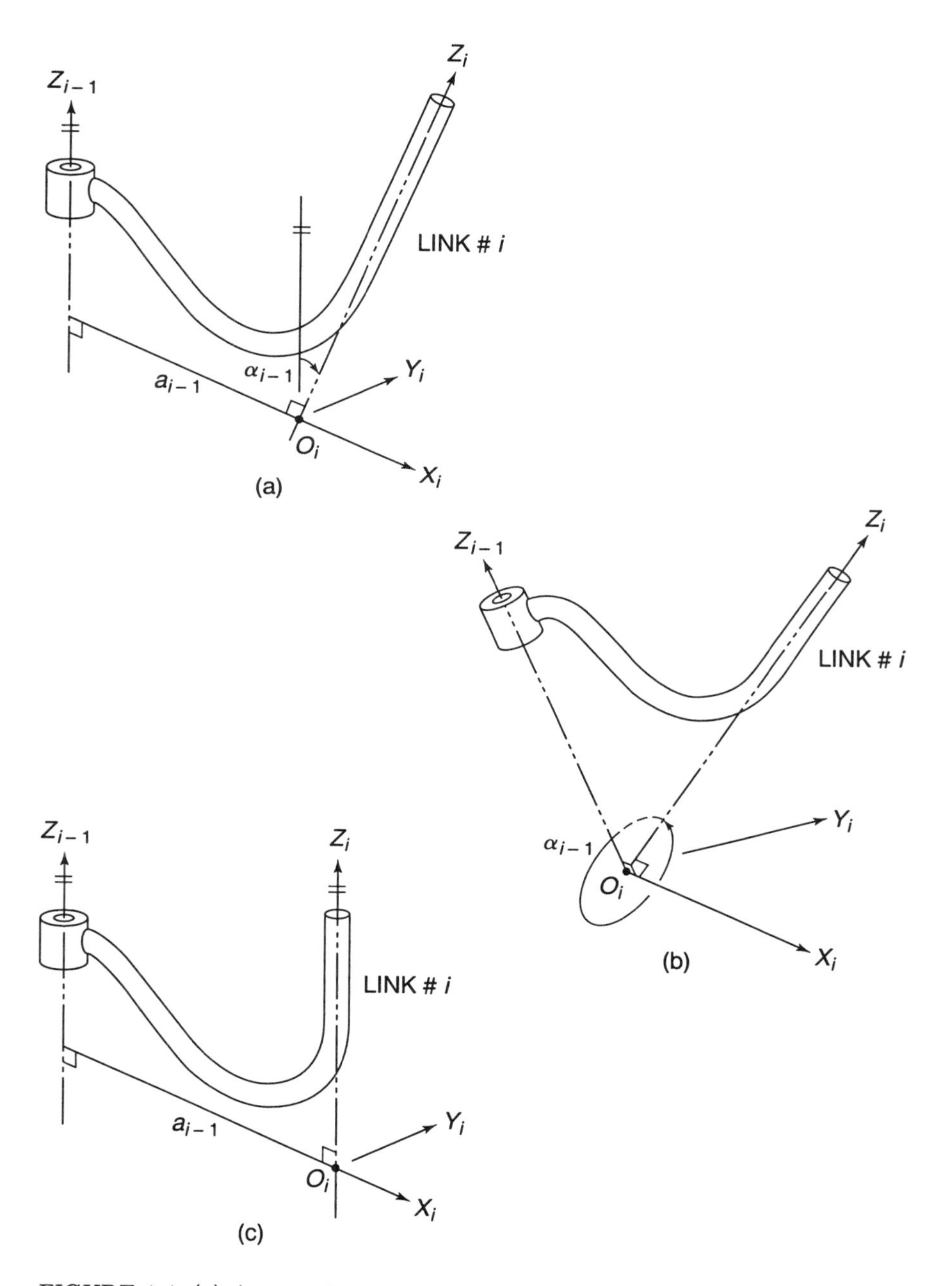

FIGURE 1.4. (a) A general spatial link—joint axes are skewed. (b) A spherical link—joint axes intersect. (c) A planar link—joint axes are parallel.

are the same. Link length a_{i-1} and link twist α_{i-1} are defined, respectively, as the shortest distance and the angle between the joint axes $(i-1)$ and i. The ith link coordinate system is attached to the ith link such that the X_i axis points along the common perpendicular to the joint axes in a direction that points from the $(i-1)$th axis to the ith axis, and the Z_i axis is directed along the ith joint axis. The value of the link length is always positive or zero. When the two link axes intersect, as in a spherical link [Figure 1.4(b)], the link length is zero, and the direction of the X_i axis can be chosen to point either way along the common perpendicular, which now passes through the point of intersection of the two joint axes. If the two joint axes are parallel, as in a planar link [Figure 1.4(c)], then the common perpendicular is not uniquely defined, and the X_i axis can be placed anywhere in accordance with the above requirements. The true angle between the joint axes can be viewed along the common perpendicular. The sign of the twist angle α_{i-1} is defined by applying the right-hand screw rule along the X_i axis. If a right-hand screw is placed along the X_i axis, then a positive twist angle α_{i-1} will move the screw in the positive X_i direction. When viewed along the $(-X_i)$ direction, the positive link twist is counterclockwise, and when viewed along the $(+X_i)$ direction, the positive link twist is clockwise.

Additional fixed parameters and variables appear when two links are connected (Figure 1.5) by means of an axial joint (R, P). Consider joint axes $(i-1)$, i, and $(i+1)$, and links i and $(i+1)$. Link coordinate systems on links i and $(i+1)$ can be defined according to the above rules. It should be noted that the X_{i+1} axis is along the common perpendicular to Z_i and Z_{i+1} axes, while the Z_i axis is along the common perpendicular to X_i and X_{i+1} axes. Thus, the X axes are along the common perpendiculars of the adjacent Z axes, while the Z axes are along the common perpendiculars of the X axes. Joint offset s_i is defined as the shortest distance between the X_i and X_{i+1} axes. It can be positive, zero, or negative. It is a fixed parameter for the revolute joint but a variable for the prismatic joint. The joint angle θ_i is defined as the angle between the X_i and X_{i+1} axes; its sign is found by applying the right-hand screw rule along the Z_i axis. It is a variable for the revolute joint but a fixed parameter for the prismatic joint. The relationship between the ith and $(i+1)$th link coordinate systems is defined completely by four fixed parameters

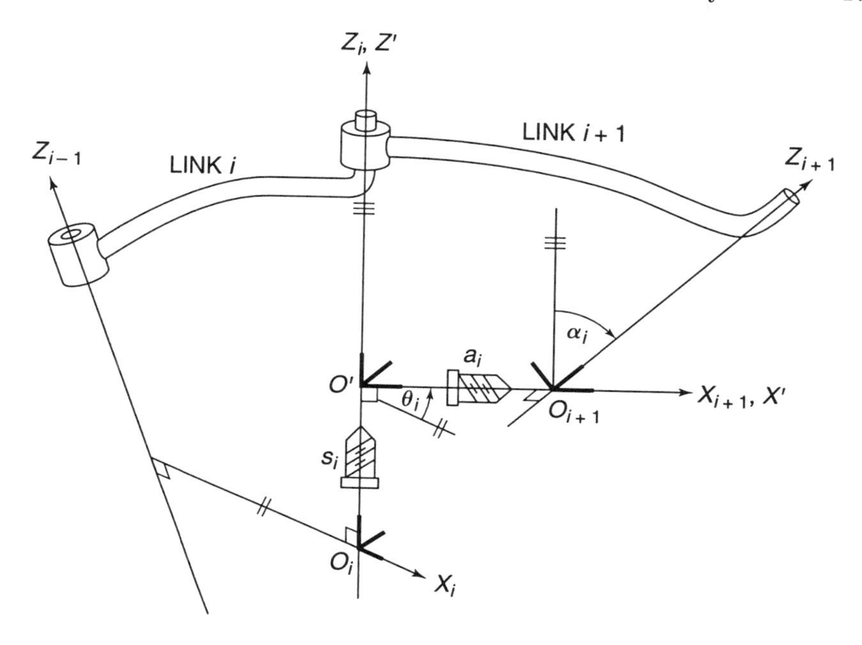

FIGURE 1.5. Adjacent links with link coordinate systems $X_iY_iZ_i$ and $X_{i+1}Y_{i+1}Z_{i+1}$. The auxiliary coordinate system $X'Y'Z'$ is also shown.

and variables:

θ_i = angle (X_i, X_{i+1}) $\qquad s_i$ = normal distance (X_i, X_{i+1})

α_i = angle (Z_i, Z_{i+1}) $\qquad a_i$ = normal distance (Z_i, Z_{i+1})

It is interesting to note the pattern of subscripts. The parameters and variables define angles or distances between two adjacent axes, but the subscript of only the first axis is retained. A double subscripted notation can be used, but according to our link numbering convention, the second subscript will usually be one more than the first subscript. These four parameters form a minimal system for describing the kinematic relationship between the adjacent link coordinate systems, and the use of explicit joint constraints is no longer required. If the Z axes are selected in a certain way, then for the general case, this system of coordinate systems and parameters/variables is also unique; exceptions occur for spherical and planar links. Although the minimal property and uniqueness are very desirable characteristics, they also make it difficult to learn this system and correctly determine the numerical values of the fixed parameters and variables.

Variations occur in the numbering of joints and links or in shifting the link coordinate system backward along the common perpendicular to the lower numbered axis. These variations do not lead to any advantages over the "DH-classic" system. In their original description in 1955, Denavit and Hartenberg did not follow the right-hand screw rule for defining positive link twist angle, but in their subsequent works, and in their famous book (1964), they have used what has been described as the "DH-classic" system here.

To derive the relation between the ith and $(i+1)$th coordinate systems, let us introduce an auxiliary (or intermediate) coordinate system $X'Y'Z'$ such that the axis X' is along the axis X_{i+1} and the axis Z' is along the axis Z_i. Then, a system coincident with the ith link system $X_iY_iZ_i$ can be moved to the auxiliary system $X'Y'Z'$ by a turn-slide displacement of rotation amount θ_i about the Z_i axis and translation amount s_i along the Z_i axis. The homogeneous transformation from the $X'Y'Z'$ system to the $X_iY_iZ_i$ system is then

$$\begin{bmatrix} X_i \\ Y_i \\ Z_i \\ 1 \end{bmatrix} = \begin{bmatrix} \cos\theta_i & -\sin\theta_i & 0 & 0 \\ \sin\theta_i & \cos\theta_i & 0 & 0 \\ 0 & 0 & 1 & s_i \\ 0 & 0 & 0 & 1 \end{bmatrix} \begin{bmatrix} X' \\ Y' \\ Z' \\ 1 \end{bmatrix}. \quad (1.10)$$

Although the general expressions will be derived later, the homogeneous transformation matrix above can be found by extending the familiar result [Eq. (1.8)] for two-dimensional rotation about the Z axis (i.e., rotation in the X–Y plane).

Next, a system coincident with the auxiliary system $X'Y'Z'$ can be moved to system $X_{i+1}Y_{i+1}Z_{i+1}$ by another turn-slide displacement of rotation amount α_i about the X' axis and translation amount a_i along the X' axis. The homogeneous transformation from the $X_{i+1}Y_{i+1}Z_{i+1}$ system to the $X'Y'Z'$ system is

$$\begin{bmatrix} X' \\ Y' \\ Z' \\ 1 \end{bmatrix} = \begin{bmatrix} 1 & 0 & 0 & a_i \\ 0 & \cos\alpha_i & -\sin\alpha_i & 0 \\ 0 & \sin\alpha_i & \cos\alpha_i & 0 \\ 0 & 0 & 0 & 1 \end{bmatrix} \begin{bmatrix} X_{i+1} \\ Y_{i+1} \\ Z_{i+1} \\ 1 \end{bmatrix}. \quad (1.11)$$

The homogeneous transformation matrix above can be found easily by extending the familiar result for two-dimensional rotation about the X axis (i.e., rotation in the Y–Z plane).

Combining this sequence of two transformations, we obtain the following homogeneous transformation from the $(i+1)$th system

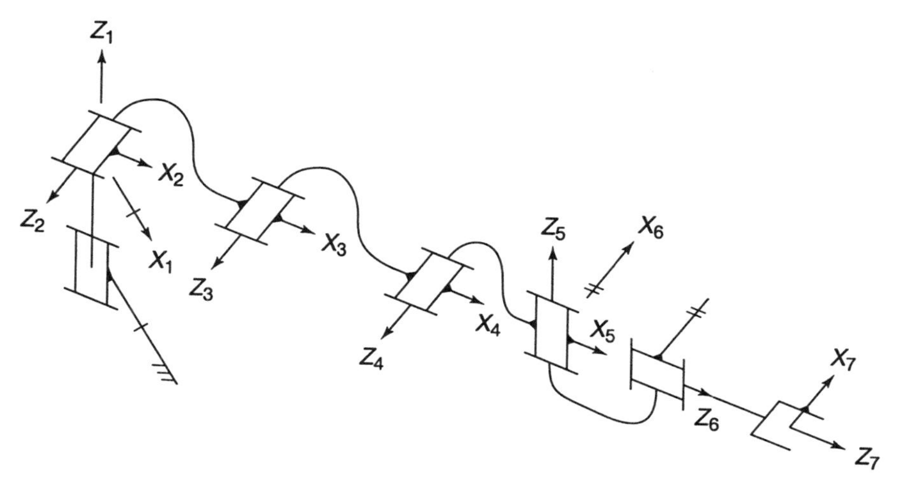

FIGURE 1.6. Kinematic diagram for Cincinnati Milacron T^3 robot with DH-system axes X_i and Z_i shown.

$X_{i+1}Y_{i+1}Z_{i+1}$ to the ith system $X_iY_iZ_i$,

$$\begin{bmatrix} X_i \\ Y_i \\ Z_i \\ 1 \end{bmatrix} = \begin{bmatrix} \cos\theta_i & -\sin\theta_i\cos\alpha_i & \sin\theta_i\sin\alpha_i & a_i\cos\theta_i \\ \sin\theta_i & \cos\theta_i\cos\alpha_i & -\cos\theta_i\sin\alpha_i & a_i\sin\theta_i \\ 0 & \sin\alpha_i & \cos\alpha_i & s_i \\ 0 & 0 & 0 & 1 \end{bmatrix} \begin{bmatrix} X_{i+1} \\ Y_{i+1} \\ Z_{i+1} \\ 1 \end{bmatrix}. \tag{1.12a}$$

The matrix of this transformation is designated simply as $[A_i]$, or more elaborately with double subscripts as $[A_{i,i+1}]$.

$$A_i = \begin{bmatrix} \cos\theta_i & -\sin\theta_i\cos\alpha_i & \sin\theta_i\sin\alpha_i & a_i\cos\theta_i \\ \sin\theta_i & \cos\theta_i\cos\alpha_i & -\cos\theta_i\sin\alpha_i & a_i\sin\theta_i \\ 0 & \sin\alpha_i & \cos\alpha_i & s_i \\ 0 & 0 & 0 & 1 \end{bmatrix}. \tag{1.12b}$$

The inverse of matrix $[A_i]$ can be written in the partitioned form as follows, omitting subscript i for convenience:

$$A_{4\times4} = \begin{bmatrix} R_{3\times3} & \mathbf{b}_{3\times1} \\ \mathbf{0} & 1 \end{bmatrix}, \qquad A^{-1} = \begin{bmatrix} R^t & -R^t\mathbf{b} \\ \mathbf{0} & 1 \end{bmatrix}. \tag{1.13}$$

Vector $\mathbf{b}$ is from the origin O_i to the origin O_{i+1}, and its components are expressed in system $X_iY_iZ_i$. It can be verified by direct multiplication of partitioned matrices that $AA^{-1} = I$.

An example of the DH coordinate systems and parameters for the Cincinnati Milacron T^3 robot is shown in Figure 1.6, where only X

TABLE 1.1. DH parameters for Cincinnati Milacron T^3.

i	s_i	α_i	a_i
1	0	90°	0
2	0	0°	a_2
3	0	0°	a_3
4	0	−90°	a_4
5	0	90°	0
6	s_6	0°	0

and Z axes are indicated. Note the attachments of axes to specific link bodies; axis X_6 belongs to link 6 but must pass through the intersection point of axes Z_5 and Z_6. The Denavit–Hartenberg parameters are shown in Table 1.1. The joint variables θ_i in the position shown in Figure 1.6 are $(\theta_1, 0°, 0°, 0°, 90°, 0°)$.

1.6 Displacements

In the active approach, the displacements of points and vectors are viewed from the fixed base system. Basic equations to represent spatial rotations and displacements will be summarized in this section. Body vectors $\mathbf{V}$ will be represented as 3×1 column vectors $(V_x, V_y, V_z)^t$, and in homogeneous representation, the point coordinates $\mathbf{P}$ will be represented as 4×1 column vectors $(\mathbf{P}^t, 1)^t$ or $(X, Y, Z, 1)^t$. For the sake of consistency in notation, the body vectors $\mathbf{V}$ can also be represented as 4×1 column vectors $(V_x, V_y, V_z, 0)^t$. The rigid body rotation can be represented in terms of a 3×3 orthogonal matrix with unit (+1) determinant, and the symbol $[R]$, or simply R, will be used for the rotation matrix. The displacement will be represented by the 4×4 matrix $[D]$, or simply D. Square brackets will be used for matrices whenever the possibility of confusion exists but will be dropped otherwise. As the body moves from position 1 to position 2, a body vector moves from $\mathbf{V}_1$ to $\mathbf{V}_2$, and a body point moves from $\mathbf{P}_1$ to $\mathbf{P}_2$; all of the components are measured with respect to the fixed base system. Then, we have the relations

$$\mathbf{V}_2 = [R_{1\to 2}]\mathbf{V}_1, \quad \begin{bmatrix} \mathbf{P}_2 \\ 1 \end{bmatrix} = [D_{1\to 2}] \begin{bmatrix} \mathbf{P}_1 \\ 1 \end{bmatrix}, \tag{1.14a}$$

where the homogeneous displacement matrix has the partitioned form

$$D_{1\to 2} = \begin{bmatrix} R_{3\times 3} & \mathbf{d} \\ \mathbf{0} & 1 \end{bmatrix}. \tag{1.15}$$

If the 4×1 column vector representation is used for body vectors, with zero in the last location, then an alternate form of Eq. (1.14a) is

$$\mathbf{V}_2 = [R_{1\to 2}]\mathbf{V}_1 \Rightarrow \begin{bmatrix} \mathbf{V}_2 \\ 0 \end{bmatrix} = [D_{1\to 2}] \begin{bmatrix} \mathbf{V}_1 \\ 0 \end{bmatrix}. \tag{1.14b}$$

Note that the homogeneous displacement matrix $[D_{1\to 2}]$ is the same in Eqs. (1.14a) and (1.14b). These expressions show the power of the active approach. The matrices $[R_{1\to 2}]$ and/or $[D_{1\to 2}]$ are determined from the given physical rotation and/or displacement. Then, the displaced body vector $\mathbf{V}$ and point $\mathbf{P}$, as seen from the fixed base system, are found directly from these expressions, knowing the initial $\mathbf{V}$ and $\mathbf{P}$. It is interesting to note that these displacement calculations can be done without knowing where the entire body is in relation to the base system and that we are able to do these calculations by using the base system only. There is no need to set up a local coordinate system on the body to do these displacement calculations (that would be the first step if the passive approach were used).

If we look at a body point that is coincident with the base system origin in body position 1, i.e., a point for which $\mathbf{P}_1 = \mathbf{0}$, then it is displaced to $\mathbf{P}_2 = \mathbf{d}$. Therefore, the physical meaning of vector $\mathbf{d}$ in Eqs. (1.6) or (1.15) is that it represents the displacement of a body point coincident with the base origin; however, as the body moves, the definition of this coincident body point changes constantly. This interpretation of vector $\mathbf{d}$ often allows us to find its components by inspection of a sketch of a simple displacement. It should be noted that the physical meaning of vector $\mathbf{b}$ in Eq. (1.13) is different from that of vector $\mathbf{d}$: vector $\mathbf{b}_i$ in the passive coordinate transformation approach represents the coordinates of the origin of the $(i+1)$th system in the ith system.

The rotation matrix is orthogonal ($R^t R = I$ or $R^{-1} = R^t$), and its determinant is $(+1)$; an orthogonal matrix with the determinant of (-1) does not represent rigid body rotation but instead represents a mirror reflection, which is quite useful in optics. The inverse displacement is

$$D_{2\to 1} = [D_{1\to 2}]^{-1} = \begin{bmatrix} R^t & -R^t\mathbf{d} \\ \mathbf{0} & 1 \end{bmatrix}. \tag{1.16}$$

Let the base system be XYZ and the body system be xyz. Then, the base coincident position (BCP) of the moving body is defined as when the systems xyz and XYZ coincide with each other [Figure 1.7(a)]. It is a hypothetical position in that it may or may not occur physically. The current position (CP) of the body is shown in Figure 1.7(b). Then, the coordinate transformation matrix of the passive approach is the special displacement matrix $[D_{\text{BCP}\to\text{CP}}]$ of the active approach. That is, the coordinate transformation matrix that changes the coordinates from the xyz system to the XYZ system in Figure 1.7(b) is also the special displacement matrix $[D_{\text{BCP}\to\text{CP}}]$ that moves the body from its base coincident position (BCP) in Figure 1.7(a) to its current (CP) position in Figure 1.7(b). This important result provides the key link between the passive and active approaches.

Turn-Slide Representation

According to Euler's theorem, any finite spherical displacement—i.e., finite spatial displacement with one fixed point—can be represented by a finite rotation about a unique axis of rotation. The rotation matrix can thus be represented as $R = R(\theta, \mathbf{u})$, where $\mathbf{u}$ is the unit vector ($\mathbf{u}^t\mathbf{u} = 1$) along the axis of rotation, and θ is the finite angle of rotation.

Chasles extended this theorem to general spatial displacements and proved that any spatial displacement can be represented as a turn-slide with respect to a unique axis in three-dimensional space; the turning is about this axis, and sliding is also along this axis (Figure 1.8). The displacement matrix can then be represented as $D = D(\theta, s, \mathbf{u}, \mathbf{Q})$, where $\mathbf{u}$ is the unit vector ($\mathbf{u}^t\mathbf{u} = 1$) along the unique axis of turn-slide, $\mathbf{Q}$ is the position vector of a point on the turn-slide axis, θ is the finite angle of rotation about the turn-slide axis, and s is the translation along the turn-slide axis. The direction and location of the turn-slide axis in space are given by $(\mathbf{u}, \mathbf{Q})$. Although the physical turn-slide axis is unique, the definition of $\mathbf{Q}$ is not because $\mathbf{Q}$ and $\mathbf{Q} + c\mathbf{u}$, c being some constant, define different points on the same turn-slide axis. One way to avoid this

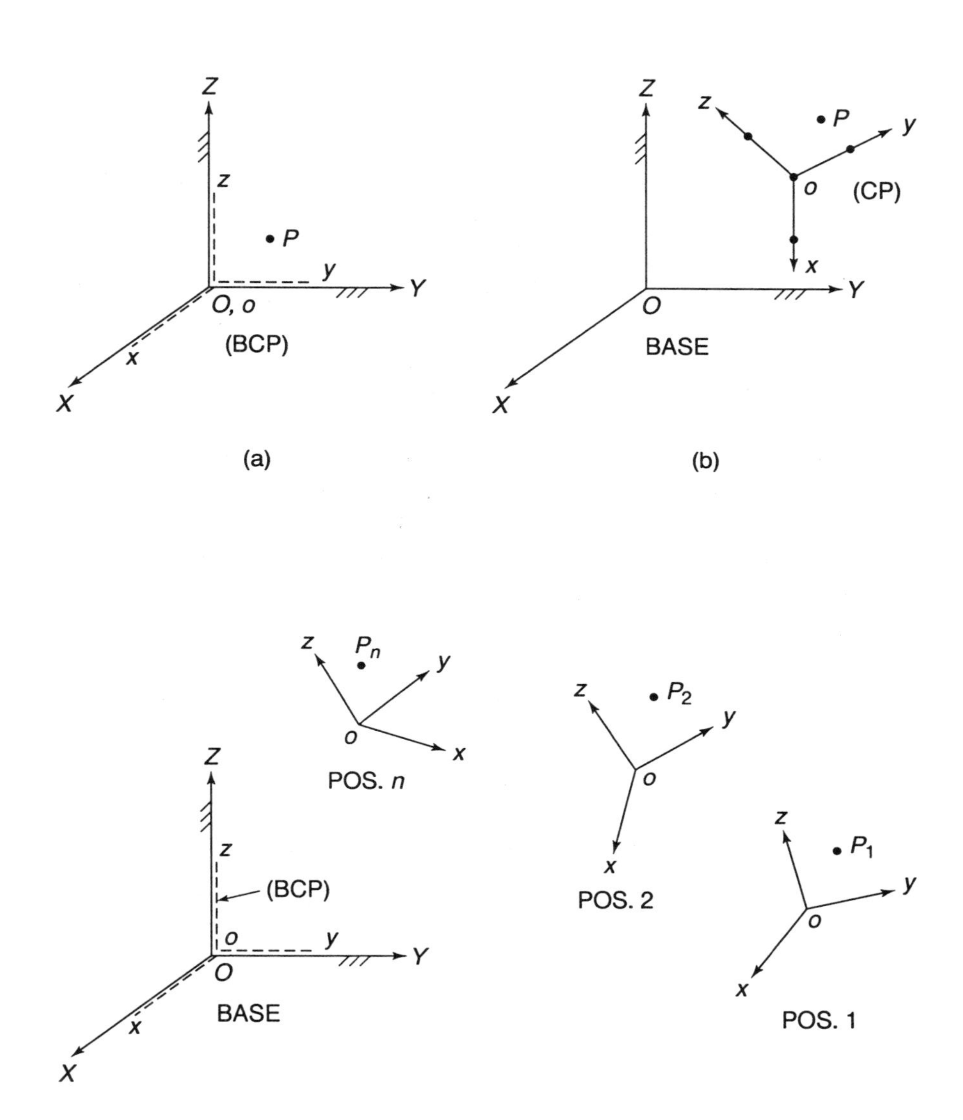

FIGURE 1.7. (a) Base coincident position (BCP) of the body system xyz. (b) Current position (CP) of the body system xyz. (c) A set of positions of the body system xyz.

problem is to impose the condition that $\mathbf{Q}$ be perpendicular to $\mathbf{u}$ ($\mathbf{Q} \perp \mathbf{u}$), i.e., $\mathbf{u}^t\mathbf{Q} = 0$. Then, the definition of $\mathbf{Q}$ becomes unique and it is directed along the unique shortest distance line from the base origin to the axis of turn-slide.

If the direction of $\mathbf{u}$ is changed by choice to $(-\mathbf{u})$, then θ and s change to $(-\theta)$ and $(-s)$, respectively, and it is still the same physical displacement. Therefore, in asserting the uniqueness of the turn-slide representation, we do not distinguish between $D(\theta, s, \mathbf{u}, \mathbf{Q})$ and $D(-\theta, -s, -\mathbf{u}, \mathbf{Q})$. This, however, should not be confused with the inverse displacement, which now is $D^{-1} = D(-\theta, -s, \mathbf{u}, \mathbf{Q})$, i.e., the rotational and translational parameters of the turn-slide are reversed in sign, but the turn-slide direction and location in space remain the same.

It may appear that the turn-slide representation of a general spatial displacement, $D = D(\theta, s, \mathbf{u}, \mathbf{Q})$, has eight parameters, but these are not independent. They satisfy two relations that have been mentioned already: $\mathbf{u}^t\mathbf{u} = 1$ and $\mathbf{u}^t\mathbf{Q} = 0$. Thus, four independent parameters are required to define the axis of turn-slide (line) in space, and an additional two parameters define the rotation and translation with respect to the turn-slide axis.

From the theory of spatial displacements, the expressions for R and $\mathbf{d}$ are

$$R = R(\theta, \mathbf{u}) = [I + U \sin\theta + U^2(1 - \cos\theta)], \tag{1.17}$$

$$\mathbf{d} = \mathbf{d}(\theta, s, \mathbf{u}, \mathbf{Q}) = s\mathbf{u} - (R - I)\mathbf{Q}, \tag{1.18}$$

where U is the 3×3 skew symmetric matrix that corresponds to the 3×1 unit vector $\mathbf{u}$, and it is defined as

$$[U] = \begin{bmatrix} 0 & -u_z & u_y \\ u_z & 0 & -u_x \\ -u_y & u_x & 0 \end{bmatrix}. \tag{1.19}$$

The expression for $\mathbf{d}$ can also be written as $\mathbf{d} = \{R(-\mathbf{Q}) - (-\mathbf{Q})\} + s\mathbf{u}$; with respect to the turn-slide, the base origin is at $(-\mathbf{Q})$ and the terms within $\{\ \}$ give the displacement of a point that is coincident with the base origin due to the rotational part of the turn-slide. When the translational part $s\mathbf{u}$ is added, we can see that $\mathbf{d}$ represents the total displacement of the point that is coincident with the base origin. It should be noted that for pure rotation (i.e., $s = 0$), $\mathbf{d}$ is not zero in general; $\mathbf{d} = 0$ only when the axis of pure rotation also passes

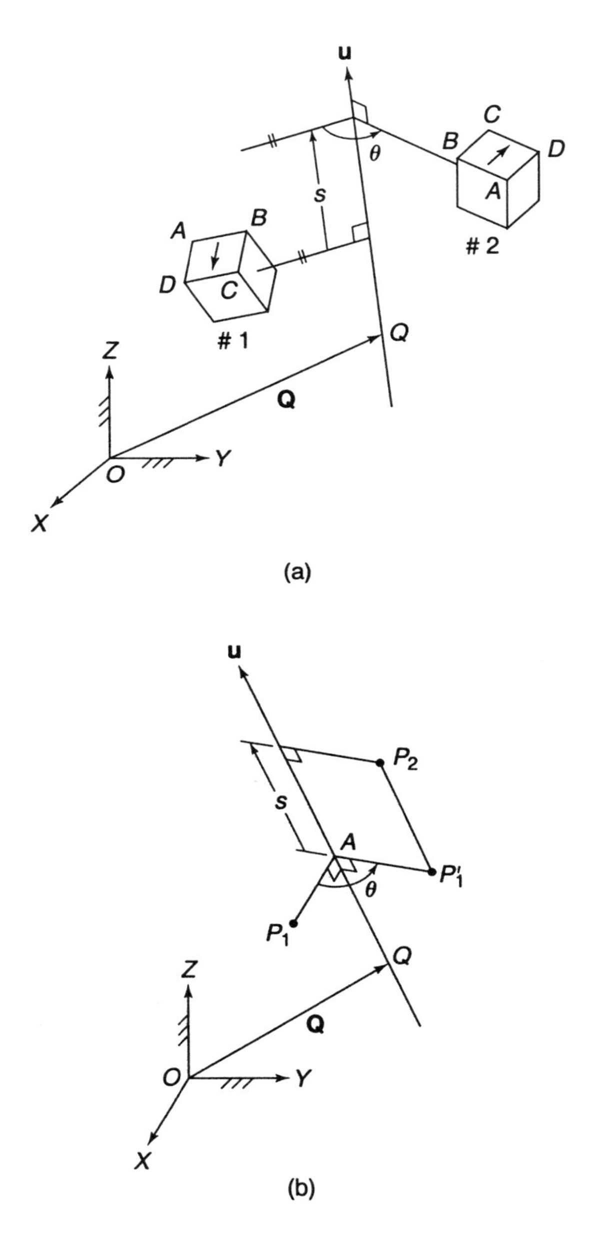

FIGURE 1.8. (a) Turn-slide representation of a general spatial displacement. (b) An initial point P_1 is rotated by angle θ about the turn-slide axis to P_1' and then translated along the turn-slide axis to P_2.

through the origin of the base system. Thus, the 3×3 principal minor of the displacement matrix D (i.e., R) contains only the rotational information; the fourth column of D contains information about both the translation and rotation.

Rotational matrix R is orthogonal, and, in general, not symmetric. However, for $\theta = \pi$ it becomes symmetric. In the theory of spatial displacements, such rotations have a special significance and are called half-turns. If θ is small, then $R \cong I + U\theta$. It can also be verified that while matrix U is skew symmetric, U^2 is symmetric, and higher powers of U have the following properties: $U^2 = \mathbf{u}\mathbf{u}^t - I_{3\times3}$, $U^3 = -U$, $U^4 = -U^2$, $U^5 = U$. It has been mentioned that $\mathbf{u}$ is a unit vector. Elements of D (or R) are very sensitive to small errors in $\mathbf{u}$. Therefore, in numerical calculations, it is recommended that the components of unit vector $\mathbf{u}$ be renormalized (i.e., divide by $\mathbf{u}^t\mathbf{u}$) to eliminate the effects of rounding and machine errors.

The elementary rotation matrices are those for rotations about the base coordinate directions **I**, **J**, **K**.

$$R(\theta, \mathbf{I}) = \begin{bmatrix} 1 & 0 & 0 \\ 0 & \cos\theta & -\sin\theta \\ 0 & \sin\theta & \cos\theta \end{bmatrix}, \tag{1.20}$$

$$R(\theta, \mathbf{J}) = \begin{bmatrix} \cos\theta & 0 & \sin\theta \\ 0 & 1 & 0 \\ -\sin\theta & 0 & \cos\theta \end{bmatrix}, \tag{1.21}$$

$$R(\theta, \mathbf{K}) = \begin{bmatrix} \cos\theta & -\sin\theta & 0 \\ \sin\theta & \cos\theta & 0 \\ 0 & 0 & 1 \end{bmatrix}. \tag{1.22}$$

Pure translation is represented as

$$D(0, s, \mathbf{u}, \mathbf{Q}) = \begin{bmatrix} 1 & 0 & 0 & su_x \\ 0 & 1 & 0 & su_y \\ 0 & 0 & 1 & su_z \\ 0 & 0 & 0 & 1 \end{bmatrix}. \tag{1.23}$$

Given a displacement matrix D (or R and $\mathbf{d}$), the parameters of the corresponding turn-slide can be found as follows. Let the elements of rotation matrix R be (r_{ij}), then

$$\begin{aligned} r_{11} &= (u_x^2 - 1)(1 - \cos\theta) + 1 \qquad \text{(diagonal elements)} \\ r_{22} &= (u_y^2 - 1)(1 - \cos\theta) + 1 \end{aligned}$$

$$\begin{aligned}
r_{33} &= (u_z^2 - 1)(1 - \cos\theta) + 1 \\
r_{12} &= u_x u_y (1 - \cos\theta) - u_z \sin\theta \qquad \text{(off-diagonal elements)} \\
r_{21} &= u_x u_y (1 - \cos\theta) + u_z \sin\theta \\
r_{13} &= u_x u_z (1 - \cos\theta) + u_y \sin\theta \\
r_{31} &= u_x u_z (1 - \cos\theta) - u_y \sin\theta \\
r_{23} &= u_y u_z (1 - \cos\theta) - u_x \sin\theta \\
r_{32} &= u_y u_z (1 - \cos\theta) + u_x \sin\theta.
\end{aligned}$$

By adding the diagonal elements, rotational angle θ can be found:

$$\theta = \text{arccosine}\left(\frac{r_{11} + r_{22} + r_{33} - 1}{2}\right). \tag{1.24}$$

From off-diagonal terms, the turn-slide axis direction can be found:

$$u_x = \frac{r_{32} - r_{23}}{2\sin\theta}, \quad u_y = \frac{r_{13} - r_{31}}{2\sin\theta}, \quad u_z = \frac{r_{21} - r_{12}}{2\sin\theta}. \tag{1.25}$$

Other elements must be used when θ approaches 0 or π; for example, the diagonal elements can be used to find the u's within ($\pm$) signs and then the off-diagonal elements used to resolve these signs. It appears that we have two possibilities for θ: θ and $(-\theta)$, which may lead to two choices for axis direction: $\mathbf{u}$ and $(-\mathbf{u})$. However, as discussed earlier, $(\theta, \mathbf{u})$ and $(-\theta, -\mathbf{u})$ represent the same physical rotation. To solve for s and $\mathbf{Q}$, use the following equations:

$$s\mathbf{u} - (R - I)\mathbf{Q} = \mathbf{d} \quad \text{and} \quad \mathbf{u}^t\mathbf{Q} = 0. \tag{1.26}$$

These vector equations represent four scalar equations in four unknowns. A simpler approach is to find s directly as $s = \mathbf{d}^t\mathbf{u}$; this can be verified after noting the relation $R^t\mathbf{u} = \mathbf{u}$ because $\mathbf{u}$ is the rotation axis. However, $(R - I)\mathbf{Q} = s\mathbf{u} - \mathbf{d}$ cannot be solved for $\mathbf{Q}$ because $(R - I)$ is a singular matrix. In fact, these three equations are not independent, and any one can be dropped and replaced with $\mathbf{u}^t\mathbf{Q} = 0$ to solve for $\mathbf{Q}$. Since we have shown that starting from any given displacement matrix D, the parameters of the unique turn-slide can be found by using the preceding equations, this can be viewed as a constructive proof of Chasles' theorem.

Example 1.1. Find the parameters $(\theta, s, \mathbf{u}, \mathbf{Q})$ of the turn-slide corresponding to the following displacement matrix:

$$D = \begin{bmatrix} 1/\sqrt{2} & 1/\sqrt{2} & 0 & 1 \\ 0 & 0 & 1 & 2 \\ 1/\sqrt{2} & -1/\sqrt{2} & 0 & 2 \\ 0 & 0 & 0 & 1 \end{bmatrix}.$$

Solution: The 3×3 rotation matrix R and 3×1 displacement vector $\mathbf{d}$ are

$$R = \begin{bmatrix} 1/\sqrt{2} & 1/\sqrt{2} & 0 \\ 0 & 0 & 1 \\ 1/\sqrt{2} & -1/\sqrt{2} & 0 \end{bmatrix}, \quad \mathbf{d} = \begin{bmatrix} 1 \\ 2 \\ 2 \end{bmatrix}.$$

Rotation angle from Eq. (1.24): $\theta = \arccos[(.707 + 0 + 0 - 1)/2] = 98.42°$.
Axis direction from Eq. (1.25): $\mathbf{u} = (-0.863, -0.357, -0.357)^t$.
From Eqs. (1.26),

$$\begin{bmatrix} -0.863 & 0.293 & -0.707 & 0 \\ -0.357 & 0 & 1 & -1 \\ -0.357 & -0.707 & 0.707 & 1 \\ 0 & -0.827 & -0.357 & -0.357 \end{bmatrix} \begin{bmatrix} s \\ Q_x \\ Q_y \\ Q_z \end{bmatrix} = \begin{bmatrix} 1 \\ 2 \\ 2 \\ 0 \end{bmatrix}.$$

Solving this linear system, we get $s = -2.292$, $\mathbf{Q} = (-0.489, 1.181, 0)^t$.

Combinations

In the active approach, a body may be moved successively through positions $1, 2, 3, \ldots, n$ [Figure 1.7(c)]. Considering a general body point P, its coordinates (or position vector) in the fixed base system can be found successively—$\mathbf{P}_2$ from $\mathbf{P}_1$ via displacement matrix $[D_{1\to2}]$, $\mathbf{P}_3$ from $\mathbf{P}_2$ via $[D_{2\to3}]$, and so on. However, $\mathbf{P}_n$ can also be found directly from $\mathbf{P}_1$ via $[D_{1\to n}]$, which is also called the resultant displacement. Equating the two results, we find

$$[D_{1\to n}] = [D_{(n-1)\to n}] \cdots [D_{2\to3}][D_{1\to2}]. \tag{1.27a}$$

Note that the first physical displacement $[D_{1\to2}]$ appears last in the equation, the second physical displacement $[D_{2\to3}]$ next to last, and so on. That is, the order in which the displacement matrices are multiplied in this equation is just the reverse of the order in which the successive physical displacements are executed. This can also be considered the rule for combining absolute displacements because all displacements are viewed from the same fixed base system. The corresponding result for combining rotations is

$$[R_{1\to n}] = [R_{(n-1)\to n}]\cdots[R_{2\to 3}][R_{1\to 2}]. \tag{1.27b}$$

Recall that the matrix multiplications are not commutative.

Example 1.2. Find the rotation matrices corresponding to the following two rotations:

(i) first rotation: 180° rotation about $\mathbf{u} = (3/5, 4/5, 0)^t$,

(ii) second rotation: 180° rotation about $\mathbf{u} = (-4/5, 3/5, 0)^t$.

Combine these into a resultant rotation, and find the corresponding parameters $(\theta, \mathbf{u})$.

Solution:

$$R_{1\to 2} = \begin{bmatrix} -7/25 & 24/25 & 0 \\ 24/25 & 7/25 & 0 \\ 0 & 0 & -1 \end{bmatrix}, \; R_{2\to 3} = \begin{bmatrix} 7/25 & -24/25 & 0 \\ -24/25 & -7/25 & 0 \\ 0 & 0 & -1 \end{bmatrix}.$$

Resultant $R_{1\to 3} = [R_{2\to 3}][R_{1\to 2}] = \begin{bmatrix} -1 & 0 & 0 \\ 0 & -1 & 0 \\ 0 & 0 & 1 \end{bmatrix}.$

Angle $\theta = \arccos(-1) = 180°$. Equations (1.25) for u's become indeterminate and cannot be used. However, from the diagonal elements of R, $u_x = u_y = 0$, $u_z = \pm 1$. The signs of u's can normally be resolved by the off-diagonal terms of R, but all of them are zero. In this special case, both $u_z = \pm 1$ are valid due to the rotation angle of $\theta = 180°$. Hence $\mathbf{u} = (0, 0, 1)^t$ or $(0, 0, -1)^t$, or $\mathbf{u} = \pm\mathbf{K}$.

Example 1.3. Consider the unit vectors $[\mathbf{I}, \mathbf{J}, \mathbf{K}]$ in the base coordinate system XYZ. A robot end-effector undergoes the following sequential rotations:

(i) rotation γ about the base $\mathbf{K}$ vector,

(ii) rotation β about the base $\mathbf{I}$ vector,

(iii) rotation α about the base $\mathbf{K}$ vector.

Combine these three rotations into a resultant rotation, and find the corresponding matrix (active viewpoint).

Solution: From Eqs. (1.20) and (1.22),

$$R_{1\to2} = R(\gamma, \mathbf{K}), \quad R_{2\to3} = R(\beta, \mathbf{I}), \quad R_{3\to4} = R(\alpha, \mathbf{K}).$$

The resultant rotation is $[R_{1\to4}] = [R_{3\to4}][R_{2\to3}][R_{1\to2}]$. After multiplications, and defining $c\alpha = \cos\alpha$, $s\alpha = \sin\alpha$, etc. for brevity,

$$R_{1\to4} = \begin{bmatrix} c\alpha\ c\gamma - s\alpha\ c\beta\ s\gamma & -c\alpha\ s\gamma - s\alpha\ c\beta\ c\gamma & s\alpha\ s\beta \\ s\alpha\ c\gamma + c\alpha\ c\beta\ s\gamma & -s\alpha\ s\gamma + c\alpha\ c\beta\ c\gamma & -c\alpha\ s\beta \\ s\beta\ s\gamma & s\beta\ c\gamma & c\beta \end{bmatrix}.$$

In the passive approach, the same point p [not shown in Figure 1.7(c)] is looked at from several different coordinate systems, say, $1, 2, 3, \ldots, n$. Let its coordinates be $\mathbf{p}_1$ (or "relative" position vector) in system 1, $\mathbf{p}_2$ in system $2, \ldots, \mathbf{p}_n$ in system n. Then, $\mathbf{p}_2$ can be changed into $\mathbf{p}_1$ via transformation matrix $[A_{12}]$, $\mathbf{p}_3$ can be changed into $\mathbf{p}_2$ via $[A_{23}]$, and so on; for this discussion, we have temporarily switched to the double subscripted notation for transformation matrices. However, $\mathbf{p}_n$ can also be changed directly into $\mathbf{p}_1$ via $[A_{1n}]$, which can be called the overall transformation. Again, equating the two results, we find

$$[A_{1n}] = [A_{12}][A_{23}] \ldots [A_{(n-1),n}]. \tag{1.28}$$

This can be considered the rule for combining relative displacements because each coordinate transformation matrix is also a special relative displacement matrix.

It should be clear from this discussion that the active and passive approaches are quite different concepts that lead to similar, yet different, sets of equations. The distinction between the two approaches is most striking in the combination rules above. The active approach is taken here as the fundamental approach, and all of the information about the passive approach can be deduced from the active approach.

Example 1.4. If a body coordinate system xyz is attached to the end-effector, with base XYZ and end-effector xyz systems being initially coincident, then write the overall coordinate transformation matrix to change the coordinates from the end-effector system to the base system after performing the following sequence of coordinate system changes:

(i) rotate xyz by angle α about the Z axis (direction $\mathbf{K}$) to get $x'y'z'$,

(ii) rotate $x'y'z'$ by angle β about (new) x' axis (direction $\mathbf{i}'$) to get $x''y''z''$,

(iii) rotate $x''y''z''$ by angle γ about (new) z'' axis (direction $\mathbf{k}''$) to get $x'''y'''z'''$.

The overall coordinate transformation is from the $x'''y'''z'''$ end-effector system to the base XYZ system (passive viewpoint).

Solution: Denote the 3×3 principal minor of $A_{i \to j}$ as $T_{i \to j}$; it transforms coordinates from system "j" to system "i". Designate the base XYZ as system 1, $x'y'z'$ as 2, $x''y''z''$ as 3, and body $x'''y'''z'''$ as 4. The overall transformation, to change coordinates from $x'''y'''z'''$ (#4) to XYZ (#1), is

$$[T_{1 \to 4}] = [T_{1 \to 2}][T_{2 \to 3}][T_{3 \to 4}]$$

and, upon substitution,

$$T_{1 \to 4} = \begin{bmatrix} \cos\alpha & -\sin\alpha & 0 \\ \sin\alpha & \cos\alpha & 0 \\ 0 & 0 & 1 \end{bmatrix} \begin{bmatrix} 1 & 0 & 0 \\ 0 & \cos\beta & -\sin\beta \\ 0 & \sin\beta & \cos\beta \end{bmatrix} \begin{bmatrix} \cos\gamma & -\sin\gamma & 0 \\ \sin\gamma & \cos\gamma & 0 \\ 0 & 0 & 1 \end{bmatrix}.$$

After matrix multiplication, and setting $c\alpha = \cos\alpha$, $s\alpha = \sin\alpha$, etc.,

$$T_{1 \to 4} = \begin{bmatrix} c\alpha\ c\gamma - s\alpha\ c\beta\ s\gamma & -c\alpha\ s\gamma - s\alpha\ c\beta\ c\gamma & s\alpha\ s\beta \\ s\alpha\ c\gamma + c\alpha\ c\beta\ s\gamma & -s\alpha\ s\gamma + c\alpha\ c\beta\ c\gamma & -c\alpha\ s\beta \\ s\beta\ s\gamma & s\beta\ c\gamma & c\beta \end{bmatrix}.$$

Although the final result is the same as in Example 1.3, the physical processes in these two examples are quite different. In Example 1.3, the rotations were about axes that were all fixed in the base system, while in Example 1.4, the rotation axes are not all fixed to the base or the body. Here, the initial (Z) axis belongs to the base system, the final (z'') axis belongs to the body system because $z'' \parallel z'''$, but the intermediate (x') axis belongs to neither. However, the intermediate axis is perpendicular to the initial and final rotation axes (i.e., $x' \perp (Z, z'')$, or vectorially, $\mathbf{i}' = \mathbf{K} \times \mathbf{k}''$).

Location of the Hand

Let us attach an orthogonal triad $\mathbf{u}_n - \mathbf{u}_t - \mathbf{u}_a$ to the hand (also gripper or end-effector) in its current position (CP); $\mathbf{u}_n = \mathbf{u}_t \times \mathbf{u}_a$. The origin of the triad is at point P. Subscripts represent normal (n), transverse (t), and axial (a) directions of the hand; transverse direction is across the jaws, and normal direction is normal to the plane of the jaws (Figure 1.9). The components of all three unit vectors are expressed in the fixed base system whose origin is O. In its base coincident position (BCP), the hand triad $\mathbf{n}-\mathbf{t}-\mathbf{a}$ coincides with base system coordinate directions $\mathbf{I}-\mathbf{J}-\mathbf{K}$, respectively. The special displacement that moves the hand triad from the BCP to CP position is

$$D_{\mathrm{BCP}\to\mathrm{CP}} = \begin{bmatrix} \mathbf{u}_n & \mathbf{u}_t & \mathbf{u}_a & \mathbf{OP} \\ 0 & 0 & 0 & 1 \end{bmatrix}. \tag{1.29}$$

The columns $\mathbf{u}_n$, $\mathbf{u}_t$, and $\mathbf{u}_a$ are, respectively, the direction cosines of the current hand unit vectors along $\mathbf{n}-\mathbf{t}-\mathbf{a}$ directions, expressed in the base system. The hand triad $\mathbf{n}-\mathbf{t}-\mathbf{a}$ can also be identified with a local hand coordinate system $x_h y_h z_h$ ($x \parallel n$, $y \parallel t$, $z \parallel a$) with origin at P. If the matrix that transforms coordinates from the hand system $x_h y_h z_h$ to the base system XYZ is A_h, then $A_h = [D_{\mathrm{BCP}\to\mathrm{CP}}]$, or

$$A_h = \begin{bmatrix} \mathbf{u}_n & \mathbf{u}_t & \mathbf{u}_a & \mathbf{OP} \\ 0 & 0 & 0 & 1 \end{bmatrix}. \tag{1.30}$$

The matrix $[A_h]$ contains the data for the current hand position and is also called the hand location matrix.

When a displacement occurs between two general positions, we will use the symbol D to represent it, but when it occurs between the BCP position and a general position, we will use the symbol A, with a letter subscript, to represent this special displacement. Matrix A is also a coordinate transformation matrix. Recall that matrix A with a numerical subscript has already been defined as the DH-coordinate transformation matrix (for example, A_2 is the second DH-matrix, across the second joint, which changes coordinates from DH-system 3 to DH-system 2).

A simple way to find the hand location matrix $[A_h]$ by measurements is to choose four noncoplanar body points in the hand and measure their body coordinates (x_i, y_i, z_i), $i = 1, \ldots, N$; $N = 4$, and base coordinates (X_i, Y_i, Z_i), $i = 1, \ldots, 4$. These data can be

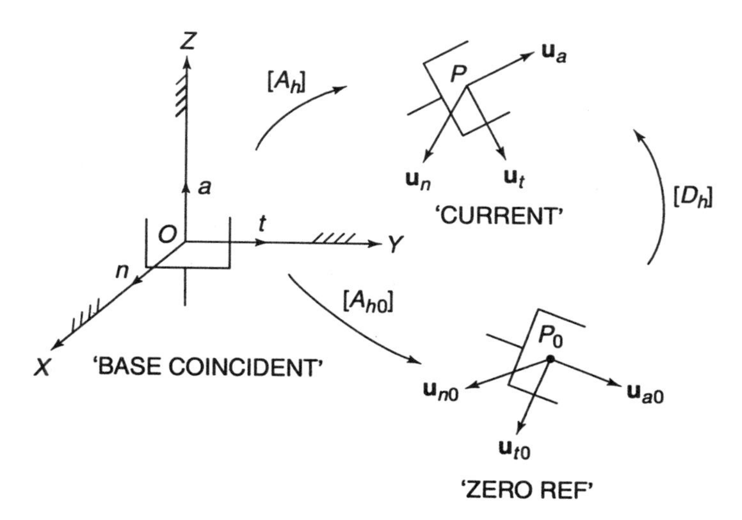

FIGURE 1.9. Three important positions of the hand: BCP—base coincident position, CP—current position, and ZRP—zero reference position. Also shown are the orthogonal triads formed by $\mathbf{n} - \mathbf{t} - \mathbf{a}$ directions.

arranged in the form $[B] = [A_h][C]$ as follows:

$$\begin{bmatrix} X_1 & X_2 & X_3 & X_4 \\ Y_1 & Y_2 & Y_3 & Y_4 \\ Z_1 & Z_2 & Z_3 & Z_4 \\ 1 & 1 & 1 & 1 \end{bmatrix} = [A_h] \begin{bmatrix} x_1 & x_2 & x_3 & x_4 \\ y_1 & y_2 & y_3 & y_4 \\ z_1 & z_2 & z_3 & z_4 \\ 1 & 1 & 1 & 1 \end{bmatrix}. \tag{1.31}$$

If the four points chosen are noncoplanar, then the matrix $[C]$ of body point coordinates on the right-hand side can be inverted, and the matrix $[A_h]$ can be found as $[A_h] = [B][C]^{-1}$. This procedure for finding $[A_h]$ is very sensitive to the errors in the measurement of the body and base coordinates. The matrix $[A_h]$ so obtained has the last row as $(0, 0, 0, 1)$, but it may not have its other theoretical properties [see Eq. (1.30)]. In classical literature, complicated formulas can be found for finding the matrix $[A_h]$ from three noncoplanar points only, which should be possible because three noncoplanar points are sufficient to fix the position of a rigid body in three-dimensional space, but these formulas make the already bad situation even worse by further magnifying the effects of measurement errors, and their use in practical work is not recommended.

If more than four points are used to reduce the influence of the measurement errors, then the matrices $[B]$ and $[C]$ containing point coordinate data become rectangular (size $4 \times N$, $N > 4$). The matrix equation $[B] = [A_h][C]$ can be solved, in the best least square sense, after postmultiplying by the transpose of the matrix $[C]$: $[B][C]^t = [A_h][C][C]^t$, where $[C][C]^t$ is a 4×4 square matrix. Computationally, the solution for $[A_h]$, or $[A_h]^t$, can be carried out by Gaussian elimination (or, equivalently, the LU-factorization of $[CC^t]$), but the result can be expressed symbolically as $[A_h] = [BC^t][CC^t]^{-1}$. Again, there is no guarantee that the resulting $[A_h]$ will have its theoretical properties. However, the last row of A_h is still found exactly as $(0, 0, 0, 1)$ because the matrices $[CC^t]$ and $[BC^t]$ have identical last rows. To improve this situation further, more sophisticated concepts from modern linear algebra may be used as follows.

Taking the transpose, reformulate the problem $[B] = [A_h][C]$ into a more familiar form for linear algebra applications:

$$[C^t]_{N\times 4}[A_h^t]_{4\times 4} = [B^t]_{N\times 4}.$$

The singular value decomposition (SVD) for the $N \times 4$ matrix $[C^t]$ can be computed as follows:

$$[C^t]_{N\times 4} = [\mathcal{Q}_1]_{N\times 4}[\mathcal{D}_{4\times 4}][\mathcal{Q}_2^t]_{4\times 4}.$$

Here, $[\mathcal{Q}_1]_{N\times 4}$ is a rectangular column-orthogonal matrix (i.e., $\mathcal{Q}_1^t\mathcal{Q}_1 = I_{4\times 4}$), $[\mathcal{D}]_{4\times 4} = [\text{diag}(\sigma_i)]$ is a square diagonal matrix with elements $\sigma_i \geq 0$, and $[\mathcal{Q}_2]_{4\times 4}$ is a square orthogonal matrix ($\mathcal{Q}_2^t\mathcal{Q}_2 = \mathcal{Q}_2\mathcal{Q}_2^t = I_{4\times 4}$). The elements σ_i are called the singular values of matrix $[C^t]_{N\times 4}$. Also define $[\mathcal{D}^+]_{4\times 4} = [\text{diag}(1/\sigma_i)]$, which is almost always the inverse $[\mathcal{D}^{-1}]$, except when some $\sigma_k = 0$, in which case the corresponding $(1/\sigma_k)$ element in $[\mathcal{D}^+]$ is set to zero and not infinity; this eliminates internal dependencies, which may be unknown a priori. Upon substitution of $[C^t]_{N\times 4}$, and simplifications with the orthogonality properties of $\mathcal{Q}_1$ (column orthogonal only) and $\mathcal{Q}_2$ (row and column orthogonal because it is a square matrix), the solution for $[A_h^t]$ is found as

$$A_h^t = [\mathcal{Q}_2]_{4\times 4}[\mathcal{D}^+]_{4\times 4}[\mathcal{Q}_1^t]_{4\times N}[B^t]_{N\times 4}$$

or, for $[A_h]$,

$$A_h = [B]_{4\times N}[\mathcal{Q}_1]_{N\times 4}[\mathcal{D}^+]_{4\times 4}[\mathcal{Q}_2^t]_{4\times 4}.$$

Although the related theory is quite complicated, the procedure above is simple to implement. Excellent library programs for singular value decomposition (SVD) are available in software packages such as EISPACK and IMSL. The least square solution of an overdetermined linear system by SVD is slow but very stable. It eliminates the nasty problems of ill-conditioning, unkown internal dependencies, and the propagation of round-off errors that commonly occur in the solution of general and large scale least square problems using the aforementioned Gaussian elimination or LU-factorization methods. Unfortunately, if the experimental data points have errors, the theoretical properties of the hand location matrix A_h are still not ensured by this procedure. In the development of robot kinematics, a bad A_h matrix can lead to large subsequent errors in computations. Therefore, the following polishing scheme can be used to bring the matrix A_h into its proper form.

Since the last row of the A_h matrix is known to be $(0, 0, 0, 1)$, any small deviations are simply the reflections of accumulated round-off errors. In the first step, the elements in the last row are set to their true values, i.e., $(0, 0, 0, 1)$. Then, the principal 3×3 minor, R_h, of the A_h matrix can be checked for orthogonality ($R_h^t R_h = R_h R_h^t = I_{3\times 3}$); again, any small deviations from $I_{3\times 3}$ indicate intrinsic data errors. These errors can be isolated through the $\mathcal{QR}$-factorization as follows:

$$[R_h]_{3\times 3} = [\mathcal{Q}]_{3\times 3}[\mathcal{R}]_{3\times 3},$$

where $[\mathcal{Q}]_{3\times 3}$ is a proper 3×3 orthogonal matrix, and $[\mathcal{R}]_{3\times 3}$ is an upper triangular matrix whose elements provide an indication of the errors of measurement that are present. If there are no experimental errors, $[\mathcal{R}]_{3\times 3} = I_{3\times 3}$. Library programs for $\mathcal{QR}$-factorization, which is related to the Gram–Schmidt orthogonalization, are also available. Then, in the second step, the 3×3 principal minor of A_h is replaced with the orthogonal matrix $[\mathcal{Q}]_{3\times 3}$. The matrix A_h now has the correct theoretical form implied by Eq. (1.30). Clearly, all of this polishing cannot eliminate intrinsic errors of experimentation, but we get an estimate of their magnitudes, and if they are unacceptable, then the precision of data gathering must be improved. Good linear algebra cannot be a substitute for bad experiments! In any event, the procedure that has been outlined does produce a good estimate of the hand location matrix A_h, which has its proper theoretical form.

Zero-Reference-Position (ZRP) Description

An alternate description of the kinematic structure of a robot manipulator [Figure 1.10(a)] is the zero-reference-position (ZRP) description. The manipulator is "frozen" in a convenient position and all joint variables (θ or s) are defined to have zero value in this zero reference position. The manipulator geometry is defined by axes directions ($\mathbf{u}_{io}$) and locations ($\mathbf{Q}_{io}$) in this position; **u**'s are unit vectors along joint axes, and **Q**'s are points that locate the joint axes. The zero reference position of the hand is specified by a reference point $\mathbf{P}_o$ and the axial and transverse unit vectors ($\mathbf{u}_{ao}, \mathbf{u}_{to}$). The joint variable of the ith revolute axis is θ_i and that of the jth prismatic axis is s_j. These ZRP data [Figure 1.10(b)] and the mathematical tools of the active approach for spatial displacements are sufficient to perform all kinds of kinematic calculations.

The zero reference position is used simply to define the geometry of the system, to provide a zero reference value for all of the variables, and to perform motion calculations, but it may not be a physically realizable position, or it may be a reachable position with some undesirable performance characteristics. This does not diminish the utility of the zero-reference-position data. The choice of the zero reference position is not unique. On the one hand, this is advantageous because the user can choose any convenient position, and the geometry description for the robot becomes very simple, even trivial, for simple industrial robots (unlike the DH parameters). On the other hand, this extreme flexibility has its price in that unless two users make identical choices, their analytical and numerical results will be different. Some users may want to use the ZRP data and the ZRP method due to their ease of learning and usage but may still want to compare their final results with those obtained by the DH data and method. This is possible at the expense of the aforementioned flexibility. Once the DH coordinate systems and parameters have been defined after a lot of hard work, the manipulator can be brought to (or redrawn in) a position where all of the joint variables in the DH representation have zero values; this zero position of the DH description can then be used as the zero reference position for the ZRP method. The ZRP data have a lot of redundant (or excess) information, and this accounts for the ease and simplicity of the ZRP description. The DH representation, on the other hand, is minimal as well as unique, in general; this is why it is more difficult to learn

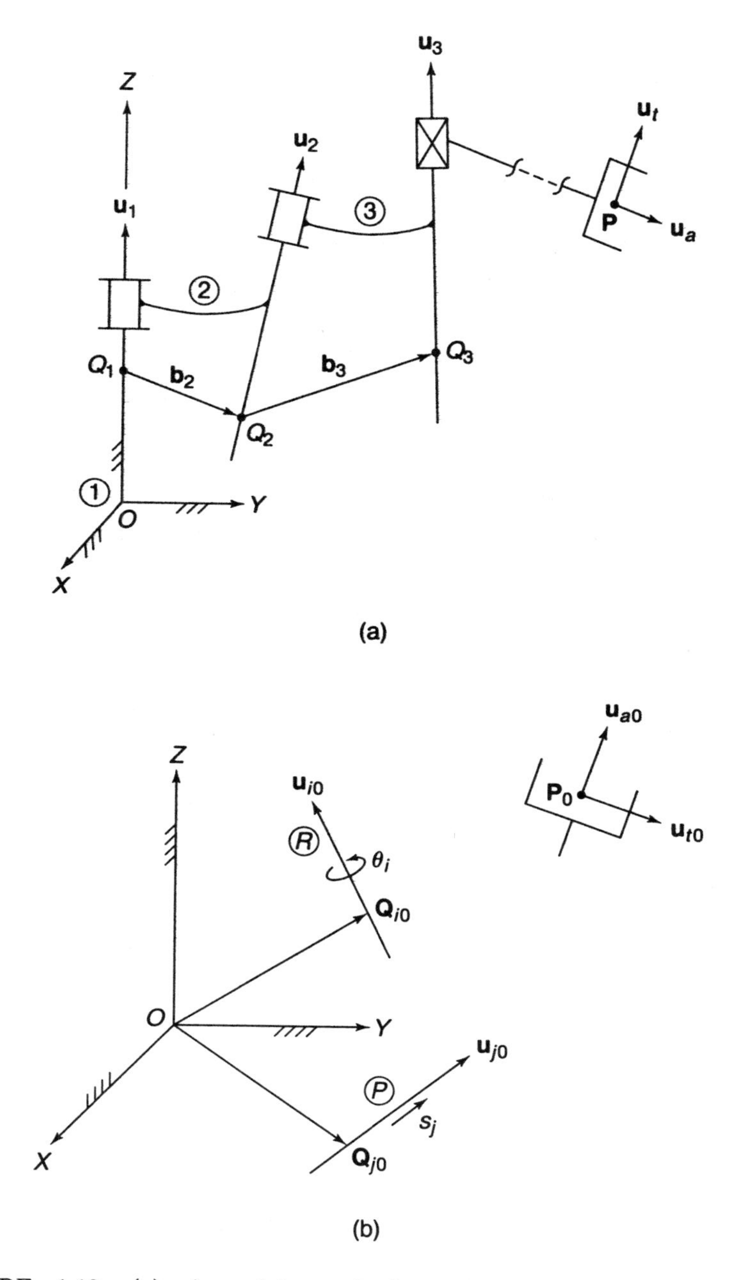

FIGURE 1.10. (a) A serial manipulator in a general position. (b) Zero-reference-position description. The locations of the joints (R or P) and the hand are "frozen" in a conveniently chosen zero reference position (ZRP), where the joint variable values are defined to be zero.

and apply correctly.

The zero-reference-position description is most suitable for open kinematic chains, which are found in serial manipulators, spatial linkages used for measuring relative motions, and some biological systems. However, it cannot be used conveniently for closed kinematic chains because an exact kinematic analysis would be required initially to obtain a consistent set of ZRP data, and that is even before the physical system could be described. This need for system analysis even before it can be properly described poses a great theoretical difficulty in applying the ZRP method to closed kinematic chains. An imperfect approach is to use an alternate description, such as the DH description, for the closed chain system, perform one accurate analysis to develop the consistent ZRP data, and then use the ZRP method. This appears to be a lot of work, but due to the inherent simplicity of the ZRP method, this approach has been tried.

Three useful positions of the hand are CP—current position, BCP—base coincident position, and ZRP—zero reference position (Figure 1.9). In the BCP position, the hand triad $\mathbf{n} - \mathbf{t} - \mathbf{a}$ is coincident with the base $\mathbf{I} - \mathbf{J} - \mathbf{K}$ directions; the triad in the CP position is $\mathbf{u}_n - \mathbf{u}_t - \mathbf{u}_a$ and in the ZRP position is $\mathbf{u}_{no} - \mathbf{u}_{to} - \mathbf{u}_{ao}$, with all unit vectors ($\mathbf{u}$'s) viewed from the base system. The triad locations are: base origin O at BCP, hand reference point P at CP, and hand reference point P_o at ZRP. The ZRP position is fixed by the ZRP data for the manipulator. Let the hand location matrices for the CP and ZRP positions be A_h and A_{ho}; these represent special BCP-to-CP and BCP-to-ZRP displacements, respectively,

$$A_h = \begin{bmatrix} \mathbf{u}_n & \mathbf{u}_t & \mathbf{u}_a & \mathbf{OP} \\ 0 & 0 & 0 & 1 \end{bmatrix}, \quad A_{ho} = \begin{bmatrix} \mathbf{u}_{no} & \mathbf{u}_{to} & \mathbf{u}_{ao} & \mathbf{OP}_o \\ 0 & 0 & 0 & 1 \end{bmatrix}. \tag{1.32}$$

Let the displacement matrix that moves the hand from the ZRP position to the current position (CP) be D_h. Then, the combination rule for displacements (active approach) leads to the following relations:

$$D_h = A_h[A_{ho}]^{-1} \quad \text{or} \quad A_h = D_h A_{ho}. \tag{1.33}$$

The relation between the D_h and A_h matrices is important because a common situation is that matrix A_h is given or known but matrix D_h is required in the calculations; or the situation could be the reverse of this. Also, the matrix A_h is used in the Denavit–Hartenberg (DH) approach, while the matrix D_h is used in the zero reference position (ZRP) approach.

1.7 Problems

1. Draw kinematic sketches for the following and indicate their degrees of freedom:

   (a) human arm-wrist system,

   (b) human fingers.

2. Two planar positions of a moving body system xoy with respect to a frame (ground) system XOY are given as $(X_{oi}, Y_{oi}; \phi_i)$: $(2'', 1''; 30°)$ and $(1'', 2''; 90°)$; $i = 1, 2$.

   (a) Sketch the three systems and find the 3×3 coordinate transformation matrices T_1 and T_2 [see Eq. (1.9)] that transform the coordinates from the body xoy system, in two positions, to the frame XOY system.

   (b) By eliminating the body coordinates (x, y) from the coordinate transform relations in part (a), find the 3×3 displacement matrix $D_{1\to2}$ [Eq. (1.6)].

   (c) The displacement matrix in part (b) above could have been found by alternate means, and note that it is independent of the choice of the moving body system xoy: it depends only upon the choice of the fixed system XOY. Describe this physical displacement in words, and also interpret it with a sketch.

3. (a) Verify the transformation equation (1.10) by sketching a view along $(-Z_i)$ in the DH-system in Figure 1.5.

   (b) Verify the transformation equation (1.11) by sketching a view along $(-X')$ in the DH-system in Figure 1.5.

   (c) For the DH-matrix A_i defined in Eqs. (1.12) and (1.13), find $[\partial A_i/\partial \theta_i][A_i]^{-1}$ and $[\partial A_i/\partial s_i][A_i]^{-1}$.

4. For a 6 dof robot manipulator with cylindric-three-roll configuration (Figure P1.1), show the DH-coordinate systems (X_i and Z_i axes only), tabulate the fixed DH-parameters, and write all of the DH-matrices A_i.

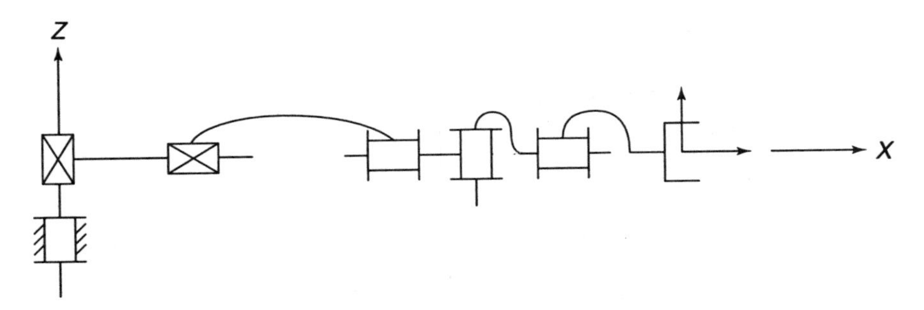

FIGURE P1.1. Cylindrical regional structure with spherical wrist equivalent for Problem 4.

5. For the Stanford Arm robot manipulator in Section 2.3, show the DH-coordinate systems (X_i and Z_i axes only), tabulate the fixed DH-parameters, and write all of the DH-matrices A_i.

6. For the 6 dof robot manipulator of Section 2.4, show the DH-coordinate systems (X_i and Z_i axes only), tabulate the fixed DH-parameters, and write all of the DH-matrices A_i.

7. For the 6 dof robot manipulator of Section 2.5, show the DH-coordinate systems (X_i and Z_i axes only), tabulate the fixed DH-parameters, and write all of the DH-matrices A_i.

8. Find displacement matrices (D) for a pure rotation of 30° about axes whose directions are $\mathbf{u} = (\sqrt{3}/4, 3/4, 1/2)^t$ but locations are: (a) $\mathbf{Q} = (0,0,0)^t$, i.e., the turn-slide axis passes through the origin, and (b) $\mathbf{Q} = (1,1,0)^t$, i.e., the turn-slide axis does not pass through the origin.

9. For the displacement matrix

$$D = \begin{bmatrix} 0 & 0 & 1 & 1 \\ 0 & 1 & 0 & 1 \\ -1 & 0 & 0 & 3 \\ 0 & 0 & 0 & 1 \end{bmatrix}$$

find the turn-slide parameters $\theta, s, \mathbf{u}, \mathbf{Q}$; use $\mathbf{u} \cdot \mathbf{Q} = 0$.

10. Find the resultant rotation matrix $R_{1\to 3}$ corresponding to the following sequence of rotations:

1st Rotation ($R_{1\to 2}$): 180° rotation about axis $\mathbf{u} = (1,0,0)^t$,

2nd Rotation ($R_{2\to3}$): 180° rotation about axis $\mathbf{u} = (1/\sqrt{2}, 1/\sqrt{2}, 0)^t$.

Find the resultant rotation parameters θ and $\mathbf{u}$.

11. (a) Consider two rotations: the first rotation is $R_{12} = R(\theta_{12}, \mathbf{u}_{12})$ and the second rotation $R_{23} = R(\theta_{23}, \mathbf{u}_{23})$. Assuming that the rotation angles θ_{ij} are small, simplify the expression for the resultant rotation matrix R_{13} by retaining only the terms of the first order in θ_{ij}.

    (b) Compare the nature of this result with that for large θ_{ij} and discuss any essential differences.

12. A sequence of four displacements is as follows:

    1st: Rotation α about the fixed X axis.

    2nd: Translation s along the fixed Z axis.

    3rd: Translation a along the fixed X axis.

    4th: Rotation θ about the fixed Z axis.

    Find the resultant displacement matrix D_{15} for this sequence of displacements.

13. Let the current displacement matrices for positions #1 and #2 be D_1 and D_2. These two positions are close, and the small displacement that can move the body from position 1 to 2 is represented by matrix D_e. The error between these two close positions can be measured in two ways: as matrix $[\nu] = [D_e - I]$ or $\Delta D = D_2 - D_1$.

    (a) Find the relation between these "displacement" error matrices $[\nu]$ and ΔD.

    (b) Let $D_e = D(\theta_e, s_e, \mathbf{u}_e, \mathbf{Q}_e)$, where θ_e and s_e are small. What is then the general simplified form of the error matrix $[\nu]$?

14. Referring to the previous problem, let the 3×3 principal minor of the error matrix $[\nu]$ be $[\varepsilon]$, and that of D_e be R_e. Then, the rotational error matrix

$$[\varepsilon] = R_e - I_{3\times3} = U_e \sin\theta_e + U_e^2(1 - \cos\theta_e).$$

Show that

$$[\varepsilon^t \varepsilon] = -2U_e^2(1 - \cos\theta_e)$$

and

$$\text{Trace}[\varepsilon^t \varepsilon] = 4(1 - \cos\theta_e).$$

The significance of this result for small angle θ_e is that the Euclidean norm of the rotational error matrix $[\varepsilon]$

$$N(\varepsilon) = \left\{ \sum_i \sum_j \varepsilon_{ij}^2 \right\}^{1/2} = \{\text{Trace}[\varepsilon^t \varepsilon]\}^{1/2} \cong 1.414\,\theta_e.$$

15. For two rotation matrices R_1 and R_2, which represent two nearby orientations, define errors $\Delta R = R_2 - R_1$ and $[\varepsilon] = R_e - I_{3\times 3}$, where $R_e = R_2(R_1)^t$. Show that these two types of "rotational" errors are related as $\Delta R = [\varepsilon](R_1)^t$ and that their Euclidean norms are equal, i.e., $N(\Delta R) = N(\varepsilon)$. *Hint*: The trace of a matrix is not affected by a similarity transformation.

16. The body coordinates of four points in a robot hand are

$$(x_i, y_i, z_i) = (3,0,1), (3,2,1), (0,2,1), (3,2,0), \ i = 1,2,3,4,$$

and the base coordinates of the same four points are, respectively,

$$(X_i, Y_i, Z_i) = (2,3,2), (2,3,0), (2,0,0), (3,3,0), \ i = 1,2,3,4.$$

(a) Find the hand location matrix A_h.

(b) If $\mathbf{u}_{ao}$ is along X, $\mathbf{u}_{no}$ along Y, and $\mathbf{u}_{to}$ along Z, and $\mathbf{P}_o$ is at the origin, find the hand displacement matrix D_h for moving the hand from the ZRP position to the current (CP) position.

17. Let the displacement matrix for a turn-slide with parameters (θ, s) with respect to axis $(\mathbf{u}_a, \mathbf{Q}_a)$ be $D_a = D(\theta, s, \mathbf{u}_a, \mathbf{Q}_a)$. The axis of the turn-slide is then moved to $(\mathbf{u}_b, \mathbf{Q}_b)$ through a displacement D_{ab}. If we want to execute a new displacement, with the same parameters (θ, s) but with respect to the new turn-slide axis $(\mathbf{u}_b, \mathbf{Q}_b)$, then the new displacement matrix $D_b = D(\theta, s, \mathbf{u}_b, \mathbf{Q}_b)$ can be found as follows (displacement similarity principle):

$$[D_b] = [D_{ab}][D_a][D_{ab}]^{-1}.$$

(a) Apply this principle to obtain the elementary turn-slide $D(\theta, s, \mathbf{J}, \mathbf{O})$ from another elementary turn-slide $D(\theta, s, \mathbf{I}, \mathbf{O})$.

(b) For rotations, this principle simplifies to

$$[R_b] = [R_{ab}][R_a][R_{ab}]^t.$$

Verify that the elementary rotation $R(\theta, \mathbf{K})$ can be found similarly from another elementary rotation $R(\theta, \mathbf{J})$.

2
Kinematic Analysis

2.1 Background

Kinematic analysis refers to position, velocity, and acceleration analysis of the manipulator. In particular, it establishes kinematic relations among the hand (end-effector) and the joint variables. The forward kinematics problem refers to the case when kinematic data (position, velocity, acceleration) are known at the joint level, but it must be found for the hand; for serial manipulators, this is a straightforward problem, as we will see later. The inverse kinematics problem refers to the situation when the kinematic data are known (or specified) for the hand but must be determined for the manipulator joints; for serial manipulators, this is a much more difficult problem than the forward kinematics problem.

The inverse position problem is highly nonlinear and difficult. The inverse velocity and acceleration problems are linear, and much simpler, once the inverse position problem has been solved. An inverse position solution is said to have a closed form if it is not iterative, i.e., the joint variable values can be found successively by using derived analytical expressions. There is only one permitted exception to the requirement of noniterative techniques, and that is the problem of determining the roots of a real polynomial in one variable.

Polynomial root finding subroutines are so highly developed that they can find all roots, real as well as complex, of a real polynomial of a reasonable degree quite reliably. In fact, for finding the roots of cubics and quartics, for which classical formulas exist, use of these subroutines is also recommended because the classical formulas are very sensitive to small errors in the coefficients. This is the only type of iterative solution that is permissible under the closed-form inverse position solution.

Among the various robot structures, the closed-form inverse position problem for manipulators with all six revolute joints is the most difficult; simplifications occur if any three successive revolute axes cointersect or are parallel. The inverse position solution also becomes simpler as the number of prismatic joints increases (but the maximum number is three). If the last three revolute joints cointersect to form a spherical wrist, then an interesting decoupling occurs in the kinematic solution (position, velocity, and acceleration). The point of cointersection (H) is called the spherical wrist center. The kinematic data for the spherical wrist center H—which are directly related to the kinematic data of the hand through rigid body kinematics—depend only upon the first three joint variables (i.e., the regional structure joints). The expressions for the position of the wrist center H provide significant simplifications in the kinematic analyses of such manipulators. In the general case, when there is no such decoupling, we must use the expressions that relate the kinematic data of the hand to all six joint variables simultaneously.

2.2 Governing Equations

Pieper–Roth Method

In a six-jointed robot manipulator, the DH-coordinate systems are defined according to the rules described previously. The base is link 1 and the coordinate system 1 (fixed) is attached to it. The hand is link 7 (or h), and the coordinate system 7 (or h) is attached to it. Six 4×4 DH matrices (A_i, $i = 1, 6$) define transformations between the adjacent links. There are 18 dimensional parameters and 6 joint variables. Consider a general point p in the three-dimensional space; it has certain coordinates in the ith DH system, which can be represented as 3×1 column vector $\mathbf{p}_i$ or the 4×1 column vector $(\mathbf{p}_i^t, 1)^t$. The transformation between system 1 and system 7 (or h) can be obtained in two ways. First, the coordinates in system 2 ($\mathbf{p}_2$) are changed into those in system 1 ($\mathbf{p}_1$) through matrix A_1 [Eq. (1.12b)], $\mathbf{p}_3$ into $\mathbf{p}_2$ through $A_2, \ldots, \mathbf{p}_7$ into $\mathbf{p}_6$ through A_6. These can be combined (passive approach rule) to get the overall transformation to change coordinates from system 7 ($\mathbf{p}_7$) to system 1 ($\mathbf{p}_1$): $A_1A_2A_3A_4A_5A_6$. Second, the hand location matrix A_h can be used directly to change coordinates from system 7 ($\mathbf{p}_7$) to system 1 ($\mathbf{p}_7$). Equating these two results [see also Eq. (1.28)], we find the following governing equation for the manipulator in matrix form,

$$A_h = A_1A_2A_3A_4A_5A_6. \tag{2.1}$$

In forward position analysis, the joint variable values are known, from which all of the A_i matrices on the right-hand side can be calculated. Then, finding the hand location matrix A_h requires the multiplication of six 4×4 matrices. This may be tedious but is rather straightforward.

In inverse position analysis, the hand motion is specified as $A_h = A_h(t)$. Each of the A_i matrices on the right-hand side now contains an unknown variable (θ_i or s_i). After multiplication, we find a dependent system of 12 highly nonlinear equations in six unknown joint variables. The dependence here is nonlinear; the equations are in fact linearly independent. In principle, only six equations can be used to find the values of the six joint variables. Three of these equations must come from matching elements $(1,4)$, $(2,4)$, and $(3,4)$; the three others must come from matching three elements of the 3×3 principal minor which are not all in the same row or column. However, multiple sets of joint variables satisfy Eq. (2.1), and to resolve all of these solution sets analytically, more than six equations, out of the twelve produced by Eq. (2.1), are needed as a rule.

The closed-form analytical solution of this problem is difficult. The term "closed form" is used in the field for those formulations that reduce the problem to the polynomial domain, i.e., at the final stage of solution, roots of a polynomial in one variable, albeit of a high order, are found numerically by using polynomial root finders or eigenvalue programs. In recent years, methods have been developed to find up to sixteen sets of real solutions that can occur for the general case of 6-R manipulators. These details are beyond the scope of this book.

If the robot arm has a spherical wrist center H, then the inverse position problem decouples and can be solved in two steps. In the first step, the coordinates of the effective wrist center point H are found in two different ways. From the geometry of the manipulator, the coordinates of point H in the DH-system 4 (i.e., $\mathbf{H}_4$) can be found. From the specification of the hand motion [$A_h(t)$ in Eq. (1.30)] and the geometry of the wrist, the coordinates of point H in the DH-system 1 (i.e., $\mathbf{H}_1$) can also be found. Then, the coordinate transformation from DH-system 4 to DH-system 1 (base) is

$$\begin{bmatrix} \mathbf{H}_1 \\ 1 \end{bmatrix} = A_1 A_2 A_3 \begin{bmatrix} \mathbf{H}_4 \\ 1 \end{bmatrix}. \tag{2.2}$$

These relations do not involve the wrist variables because the wrist joint rotations do not affect the position of the wrist center point H. Three scalar equations that result from this can be solved for the three unknown joint variables of the regional structure. This completes the first step.

In the second step, the rotational part (principal 3×3 minor) of the governing equation (2.1) is rearranged as follows:

$$R_4 R_5 R_6 = [R_1 R_2 R_3]^{-1} R_h = [R_1 R_2 R_3]^t R_h. \tag{2.3}$$

The right-hand side is known at this stage of solution, and the three unknown wrist angles θ_4, θ_5, θ_6 can be found from equations corresponding to a judicious selection of up to five elements that belong to either two rows, or one row and a column from Eqs. (2.3). The reader may wonder why it is necessary to consider more than three equations to determine the three unknown wrist angles. The reason is that a spherical wrist has two sets of joint angles that can satisfy Eq. (2.3). The extra equations are used to resolve this matter.

The equations necessary for velocity and acceleration analyses can be derived by differentiating the governing equation (2.1).

Zero-Reference-Position (ZRP) Method

The ZRP data for joint axes are $(\mathbf{u}_{io}, \mathbf{Q}_{io})$, $i = 1, 6$. A joint displacement matrix D_i can be defined as

$$D_i = D(\theta_i, s_i, \mathbf{u}_{io}, \mathbf{Q}_{io}). \tag{2.4}$$

For revolute joints, θ_i is variable and $s_i = 0$; for prismatic joints, s_i is variable and $\theta_i = 0$. Note that while the current values of joint variables are used to find D_i, the ZRP axes data $(\mathbf{u}_{io}, \mathbf{Q}_{io})$ are always used no matter where the robot is, i.e., the axis data used for D_i are not updated.

In a manipulator, for a given set of joint variable values, the location of the hand is unique, i.e., the forward position problem has a unique solution (this is not true for the inverse position problem). It does not matter how these joint variable values are achieved—by changing them one at a time or by simultaneous changes—but once the same set of joint values is reached, the same position of the hand will be achieved. If the joint variable changes are to be made one at a time, even then there are a large number of ways to make the complete set of changes. If we wish to use the D_i matrices to represent joint changes, then we must look for a sequence of joint variable changes that does not disturb those joint axes whose variables have not yet been changed. There is only one sequence, from the hand to the base, that achieves this. For example, when joint variable 6 is changed by matrix D_6, joint axes 1 to 5 are not affected; next, when joint variable 5 is changed by matrix D_5, joint axes 1 to 4 are not affected—the fact that joint 6 will be affected is not material because we have already set it to its desired value, and so on. Thus, displacements $D_6, D_5, \ldots, D_1$, constitute a sequence of successive displacements that will change all of the joint values from their zero values to the desired current value. These displacements can be combined by using the active approach rule [see Eq. (1.27)]. During this process, the hand also moves from its ZRP position to the current position (CP), which is the direct effect of the hand displacement matrix $D_h = D_{\mathrm{ZRP}\to\mathrm{CP}}$ [see Figure 1.9(a) and Eq. (1.33)]. Equating the two results, we find the following governing equation:

$$D_h = D_1 D_2 D_3 D_4 D_5 D_6. \tag{2.5}$$

Although similar in appearance to Eq. (2.1), this equation is quite unlike Eq. (2.1). The concepts involved are totally different. In Eq. (2.5), the base coordinate system is the only coordinate system used. The hand displacement matrix $D_h = A_h A_{ho}^{-1}$. The joint matrices D_i use only the ZRP axis data $(\mathbf{u}_{io}, \mathbf{Q}_{io})$, which are not updated even as the manipulator moves. The only variables on the right-hand side of Eq. (2.5) are the current values of the joint variables.

The remainder of this subsection may appear to be very similar to the previous subsection, but this is done intentionally so that each subsection can be read independently.

In forward position analysis, the joint variable values are known, from which all of the D_i matrices on the right-hand side can be calculated. Then, finding the hand displacement matrix D_h requires the multiplication of six 4×4 matrices. This may be tedious, but it is rather straightforward.

In inverse position analysis, the hand motion is specified as $D_h = D_h(t)$. Each of the D_i matrices on the right-hand side now contains an unknown variable (θ_i or s_i). After multiplication, we find a dependent system of 12 highly nonlinear equations in six unknown joint variables. The closed-form analytical solution of this system of equations is difficult.

If the robot arm has a spherical wrist center H, then the inverse position problem can be solved in two steps. From the geometry of the manipulator, the coordinates of point H in the zero reference position ($\mathbf{H}_o$) can be found. From the specification of hand motion $[D_h(t)]$ and the geometry of the wrist, the coordinates of point H in the base system ($\mathbf{H}$) can also be found. Then, the displacement of the wrist center from $\mathbf{H}_o$ (ZRP) to $\mathbf{H}$ (CP) in view of the aforementioned discussion is

$$\begin{bmatrix} \mathbf{H} \\ 1 \end{bmatrix} = D_1 D_2 D_3 \begin{bmatrix} \mathbf{H}_o \\ 1 \end{bmatrix}. \tag{2.6}$$

Three scalar equations that result from this can be solved for the three unknown joint variables of the regional structure. This then completes the first step of the ZRP method.

In the second step, the rotational part (principal 3×3 minor) of the governing equation (2.5) is rearranged as follows:

$$R_4 R_5 R_6 = [R_1 R_2 R_3]^{-1} R_h = [R_1 R_2 R_3]^t R_h. \tag{2.7}$$

Although similar in appearance to Eq. (2.3) of the previous section, this equation in the ZRP method is different in detail. The right-hand side is known at this stage, and the three unknown wrist angles θ_4, θ_5, θ_6 can be found from equations corresponding to a judicious selection of two rows, or one row and a column from Eq. (2.7).

If the axial direction of the hand is coaxial with wrist joint 6, then for Eq. (2.6),

$$\mathbf{H} = \mathbf{P} - h\mathbf{u}_a, \tag{2.8}$$

where the length of the hand $h = HP = H_o P_o$. Also, instead of using Eq. (2.7), we proceed as follows. The rotation of the axial direction of the hand from $\mathbf{u}_{ao}$ to $\mathbf{u}_a$ is not affected by joint 6, and we can write

$$\mathbf{u}_a = [R_1 R_2 R_3 R_4 R_5]\mathbf{u}_{ao} \tag{2.9}$$

or, after defining $\mathbf{v}_a = [R_1 R_2 R_3]^t \mathbf{u}_a$,

$$\mathbf{v}_a = [R_1 R_2 R_3]^t \mathbf{u}_a = [R_4 R_5]\mathbf{u}_{ao}. \tag{2.10}$$

At this stage, $\mathbf{v}_a$ is known, and unknown joint angles θ_4 and θ_5 can be found from the three scalar equations that result from Eq. (2.10).

Then, the rotation of the transverse direction of the hand from $\mathbf{u}_{to}$ to $\mathbf{u}_t$ gives

$$\mathbf{u}_t = [R_1 R_2 R_3 R_4 R_5 R_6]\mathbf{u}_{to} \tag{2.11}$$

or, after defining $\mathbf{w}_t = [R_1 R_2 R_3 R_4 R_5]^t \mathbf{u}_t$,

$$\mathbf{w}_t = [R_4 R_5]^t [R_1 R_2 R_3]^t \mathbf{u}_t = [R_6]\mathbf{u}_{to}. \tag{2.12}$$

Since $\mathbf{w}_t$ is now known, the unknown joint angle θ_6 can be found from this set of equations.

The equations necessary for velocity and acceleration analyses can be derived by differentiating the governing equation (2.5).

We will now show the details of the ZRP method by several examples. One of these examples will demonstrate that identical results can be obtained from the Pieper–Roth method and the ZRP method by selecting a special zero reference position.

Two trigonometric tricks will be found to be very useful in analysis. The inverse trigonometric functions are double valued: $\theta = \text{arcsine}(y)$ leads to angles θ and $\pi - \theta$; $\theta = \arccos(x)$ leads to angles θ and $-\theta$; $\theta = \arctan(z)$ leads to angles θ and $\pi + \theta$. When the sine and cosine of an angle are known, then the unique value of the angle defined by them can be found simply as follows:

$$\cos\theta = x,\ \sin\theta = y \quad \Rightarrow \quad \theta = 2\arctan\left(\frac{y}{1+x}\right),\ x \neq -1. \tag{2.13}$$

If $x = -1$, then this expression cannot be used, but in that case, $\theta = \pi$. To see the usefulness of this trick, consider the example when $\cos\theta = -3/5$ and $\sin\theta = 4/5$. If we find $\tan\theta = -4/3$, then the arctan function in the calculator gives $\theta = -53.13°$; arccos gives $\theta = 126.87°$ and arcsine gives $\theta = +53.13°$. We also know that there is only one correct value, so what could be wrong? Error in the math processor? Moody calculator? No! The problem is that inverse trigonometric functions are double-valued, and the calculator (or, a computer) gives only their principal values, so the correct value for the problem at hand must be figured out by the user. Geometrically, this means that the right quadrant for the angle must be determined. The formula in Eq. (2.13) resolves this problem neatly: $\theta = 2\arctan(2) = 126.87°$, and that is the unique correct value. The reader may find this puzzling because arctan is still a double-valued function. However, while the two values for the arctan function are (angle) and (angle + 180°), the two values of (2 arctan) function are (angle) and (angle + 360°), and the latter value is redundant. So, the (2 arctan) function is a single-valued function in the 0° to 360°, or −180° to 180°, range.

Another useful trick is to convert a trigonometric expression of the form $a\sin\theta + b\cos\theta + c = 0$ into a quadratic in $t = \tan(\theta/2)$, and solve for two values of θ as follows:

$$t = \tan\left(\frac{\theta}{2}\right), \quad \sin\theta = \left(\frac{2t}{1+t^2}\right), \quad \cos\theta = \left(\frac{1-t^2}{1+t^2}\right)$$

$$\Rightarrow \theta = 2\arctan\left(\frac{-a \pm \sqrt{a^2+b^2-c^2}}{c-b}\right). \tag{2.14}$$

Again, note that the (2 arctan) function is single-valued, and the two values of θ so obtained are the correct values.

2.3 Stanford Arm Manipulator

This manipulator has $R \perp R \perp P$ regional structure and a spherical wrist. There is a shoulder ($R \perp R$) and a wrist (cointersecting $R \perp R \perp R$), but there is no elbow. It does not have the capability to bend around obstacles. A small offset along joint axis 2 is provided to clear the base column during the arm motion; without this offset, the regional structure would become the model of the spherical coordinate system [Figure 1.3(a)]. The prismatic joint (3rd) gives the arm a long reach, but it also contributes to friction and accuracy problems. Starting in the late 1960s, this type of arm served as a test bed for the hand–eye project at the Stanford Artificial Intelligence Laboratory for many years, hence the name.

The zero reference position for the Stanford Arm is shown in Figure 2.1, and the ZRP data are shown in Table 2.1. The figure is

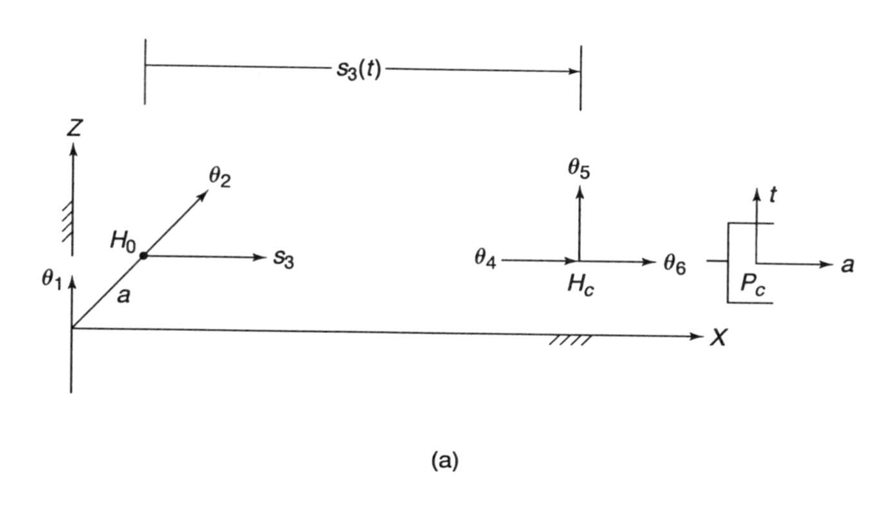

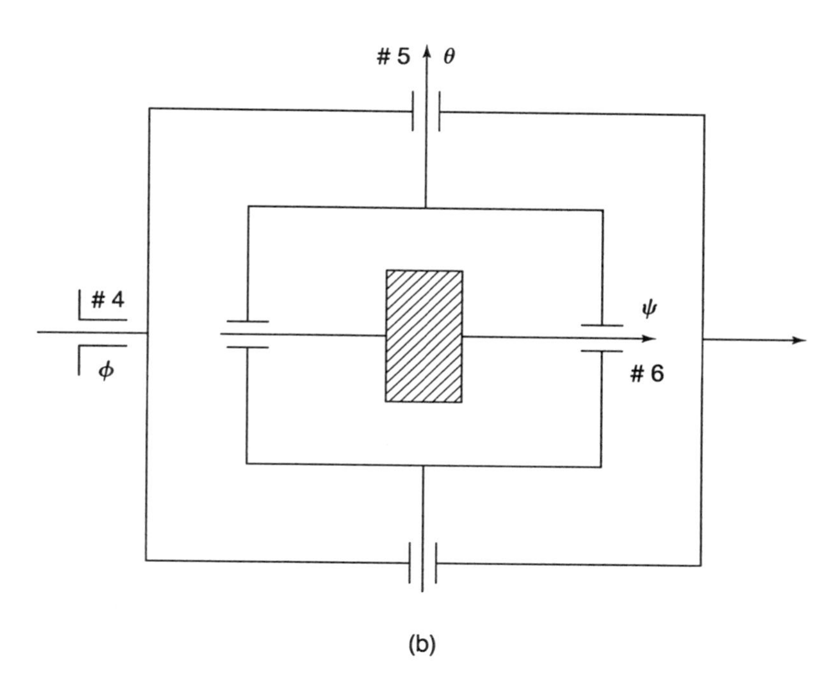

FIGURE 2.1. (a) ZRP data for the Stanford Arm manipulator. (b) Identification of spherical-wrist angles θ_4, θ_5, θ_6 with the familiar Euler angles ϕ, θ, and ψ for the chosen zero reference position.

TABLE 2.1. ZRP data for the Stanford Arm.

Axis No.	Axis Type	Dir. $\mathbf{u}_{io}$	Loc. $\mathbf{Q}_{io}$
1	R	$0, 0, 1$	$0, 0, 0$
2	R	$0, 1, 0$	$0, 0, 0$
3	P	$1, 0, 0$	$0, a, 0$
4	R	$1, 0, 0$	$0, a, 0$
5	R	$0, 0, 1$	$0, a, 0$
6	R	$1, 0, 0$	$0, a, 0$

Hand: $\mathbf{u}_{ao} = (1, 0, 0)$, $\mathbf{u}_{to} = (0, 0, 1)$, $\mathbf{P}_o = (h, a, 0)$.

a bare bone sketch, which shows the joint axis direction ($\mathbf{u}_{io}$) and locations ($\mathbf{Q}_{io}$). Because any point in the joint axis can be used to define $\mathbf{Q}_{io}$, the base origin is used to locate joint axes 1 and 2. The wrist and the hand are not shown at their ZRP positions but in their current positions, for the sake of clarity. At the ZRP position, the wrist center H is at point H_o: $(0, a, 0)$; $s_3 = 0$ then, and the hand reference point P is at P_o: $(h, a, 0)$ such that the length of the hand is $h = HP = H_oP_o$. The point H_o (also the shoulder point) thus locates joint axes 3, 4, 5, and 6. The current value of the third joint variable can be seen in Figure 2.1 as $s_3 = H_oH_c$.

The problem is: Given the current hand location as $\mathbf{u}_a$, $\mathbf{u}_t$, and $\mathbf{P}$, find all of the joint variables.

Regional Structure Solution

The current position (CP) of the spherical wrist center H: (H_x, H_y, H_z) is obtained as $\mathbf{H} = \mathbf{P} - h\mathbf{u}_a$, and its zero reference position (ZRP) is $H_o = (0, a, 0)$. Displacement of $\mathbf{H}_o$ to $\mathbf{H}$ is according to Eq. (2.6), where

$$D_1 = D(\theta_1, 0, \mathbf{K}, \mathbf{O}), \ D_2 = D(\theta_2, 0, \mathbf{J}, \mathbf{O}), \ D_3 = D(0, s_3, \mathbf{I}, \mathbf{H}_o)$$

are either obtained from Eqs. (1.17) and (1.18) or can be written by inspection in simple cases such as above [recall Eqs. (1.20)–(1.23) and the physical meaning of vector **d**].

$$D_1 = \begin{bmatrix} c\theta_1 & -s\theta_1 & 0 & 0 \\ s\theta_1 & c\theta_1 & 0 & 0 \\ 0 & 0 & 1 & 0 \\ 0 & 0 & 0 & 1 \end{bmatrix}, \quad D_2 = \begin{bmatrix} c\theta_2 & 0 & s\theta_2 & 0 \\ 0 & 1 & 0 & 0 \\ -s\theta_2 & 0 & c\theta_2 & 0 \\ 0 & 0 & 0 & 1 \end{bmatrix},$$

$$D_3 = \begin{bmatrix} 1 & 0 & 0 & s_3 \\ 0 & 1 & 0 & 0 \\ 0 & 0 & 1 & 0 \\ 0 & 0 & 0 & 1 \end{bmatrix}.$$

After expanding Eq. (2.6), we find

$$\begin{aligned} H_x &= s_3 \cos\theta_1 \cos\theta_2 - a \sin\theta_1 \\ H_y &= s_3 \sin\theta_1 \cos\theta_2 + a \cos\theta_1 \\ H_z &= -s_3 \sin\theta_2. \end{aligned}$$

Define $H_Q = \pm[H_x^2 + H_y^2 - a^2]^{1/2}$, and note that there are two possible choices. Then,

$$H_Q = s_3 \cos\theta_2.$$

Because of the unknown s_3, we do not have known values for $\sin\theta_2$ and $\cos\theta_2$, and the trick in Eq. (2.13) cannot be used yet. However, by dividing, we get an expression for $\tan\theta_2$. The solution for θ_2 (note two possible choices) and s_3 is

$$\theta_2 = \arctan(-H_z/H_Q); \quad \text{also, } (\theta_2 + 180^\circ)$$

$$s_3 = -H_z/\sin\theta_2.$$

To solve for θ_1, rearrange the expressions for H_x and H_y as (note $H_Q = s_3 \cos\theta_2$)

$$\begin{bmatrix} H_Q & -a \\ a & H_Q \end{bmatrix} \begin{bmatrix} \cos\theta_1 \\ \sin\theta_1 \end{bmatrix} = \begin{bmatrix} H_x \\ H_y \end{bmatrix}.$$

Although $\cos\theta_1$ and $\sin\theta_1$ are functionally dependent, they are linearly independent. After solving for them, we use Eq. (2.13) to solve for θ_1:

$$\theta_1 = 2\arctan\left(\frac{-aH_x + H_y H_Q}{a^2 + H_Q^2 + aH_y + H_x H_Q}\right).$$

The four sets of solutions, which are found by using all possible choices indicated above, have the following pattern:

$$\begin{array}{ccc} \theta_1 & \theta_2 & s_3, \\ \theta_1' & \pi - \theta_2 & s_3, \\ \theta_1 & \pi + \theta_2 & -s_3 \\ \theta_1' & -\theta_2 & -s_3. \end{array}$$

Clearly, the inverse position solution is not unique. In this case, there are four distinct sets of joint variables that correspond to the same current position of the wrist center H. As a practical matter, only two of these solution sets will be physically realizable because it would be impossible for the hand to slide through the hardware associated with the prismatic joint construction in order to change the sign of the third joint variable s_3.

The conditions that cause the otherwise distinct joint values to coalesce will later be found to cause singularities in the velocity analysis. Therefore, useful physical insights into manipulator singularities can be found from the inverse position solution. The regional structure solutions for Stanford Arm coalesce when:

(i) $H_Q = 0$, i.e., $H_x^2 + H_y^2 = a^2$. This occurs whenever a vertical projection of the wrist center H lies on the shoulder offset circle of radius a. Physically, this happens when the arm is either up or down. The wrist center has only 2 dof (instead of the normal 3) because in the plane that contains joint axes 1 and 2, its horizontal motion is not possible.

(ii) The two values of θ_2 become the same. This does not seem possible for any definite value of the angle because the two values are θ_2 and $(\pi+\theta_2)$. However, θ_2 becomes indeterminate when $s_3 = 0$. This is not a new situation because it is covered under (i) as a special case of the arm being up or down when the wrist center H moves to the shoulder point H_o.

Example 2.1. Find the four regional structure solutions for the Stanford Arm robot when the shoulder offset $a = 5''$, and the wrist center H is located at $(10'', 10'', 10'')$.

Solution:

$$H_Q = \pm[H_x^2 + H_y^2 - a^2]^{1/2} = \pm 13.2288''.$$

For $H_Q = +13.2288$", there are two values of angle θ_2,

$$\theta_2 = -37.087°, \quad 142.913°.$$

The corresponding values for s_3 are

$$s_3 = 16.5832'', \quad -16.5832''.$$

The first joint angle for each pair of (θ_2, s_3) is 24.295°, i.e.,

$$\theta_1 = 24.295°, \quad 24.295°.$$

Repeating the procedure for $H_Q = -13.2288''$, we get

$$\theta_2 = 37.087°, \qquad 217.087°$$

$$s_3 = -16.5832'', \qquad 16.5832''$$

$$\theta_1 = -114.296°, \qquad -114.296°.$$

Wrist Solution

In Eq. (2.10), substitute $\mathbf{u}_{ao} = (1, 0, 0)^t$ and

$$R_1 = R(\theta_1, \mathbf{K}),\ R_2 = R(\theta_2, \mathbf{J}),\ R_3 = I,$$

$$R_4 = R(\theta_4, \mathbf{I}),\ R_5 = R(\theta_5, \mathbf{K}),$$

where R_1, R_2, and R_3 are the 3×3 principal minors of D_1, D_2, and D_3 defined previously, and

$$R_4 = \begin{bmatrix} 1 & 0 & 0 \\ 0 & c\theta_4 & -s\theta_4 \\ 0 & s\theta_4 & c\theta_4 \end{bmatrix}, \quad R_5 = \begin{bmatrix} c\theta_5 & -s\theta_5 & 0 \\ s\theta_5 & c\theta_5 & 0 \\ 0 & 0 & 1 \end{bmatrix}.$$

Then,

$$v_{ax} = \cos\theta_5,\ v_y = \cos\theta_4 \sin\theta_5,\ v_{az} = \sin\theta_4 \sin\theta_5,$$

where $\mathbf{v}_a = [R_1 R_2 R_3]^t \mathbf{u}_a$ is known. The (arccos) function gives two possible choices for θ_5, and then the value of θ_4 is found by using Eq. (2.13).

$$\theta_5 = \arccos(v_{ax}),\ \text{also } (-\theta_5); \quad \theta_4 = 2\arctan\left(\frac{v_{az}}{\sin\theta_5 + v_{ay}}\right).$$

Use Eq. (2.12), $\mathbf{u}_{to} = (0,0,1)^t$, $R_6 = R(\theta_6, \mathbf{I})$, i.e.,

$$R_6 = \begin{bmatrix} 1 & 0 & 0 \\ 0 & c\theta_6 & -s\theta_6 \\ 0 & s\theta_6 & c\theta_6 \end{bmatrix},$$

and Eq. (2.13) to find θ_6,

$$\theta_6 = 2\arctan\left(\frac{-w_{ty}}{1 + w_{tz}}\right),$$

where $\mathbf{w}_t = [R_1 R_2 R_3 R_4 R_5]^t$ is known. The pattern of the two sets of wrist solutions is $(\theta_4, \theta_5, \theta_6)$ and $(\pi + \theta_4, -\theta_5, \pi + \theta_6)$. The wrist solutions coalesce when $\theta_5 = 0$ or π, i.e., the three wrist axes become coplanar. The wrist then has only 2 dof because hand rotation in the plane containing joint axes 4, 5, and 6 is not possible.

If the wrist structure ZRP data are similarly defined for other robots, then the above wrist solution will be valid except that the values of $\mathbf{v}_a$ [Eq. (2.10)] and $\mathbf{w}_t$ [Eq. (2.12)] will be appropriately modified.

Considering all possible choices in the solution procedure, the Stanford Arm manipulator theoretically has eight $(= 4 \times 2)$ sets of solutions, but only four $(= 2 \times 2)$ of these are practically possible.

It should be emphasized that the ZRP used contained a singular (undesirable) configuration for the regional structure and for the wrist structure as well. This means only that in some operations of the robot we should avoid $s_3 = 0$ and $\theta_5 = 0$, but this ZRP turned out to be very good for the analytical derivations.

2.4 Manipulator Case 2

This manipulator has $R \perp R \parallel R$ regional structure and cointersecting $R \perp R \perp R$ wrist. Joint 3 forms an elbow, and this manipulator has the capability to bend around obstacles. Joint axes 1 and 2 do not intersect, but the distance between them is relatively small; these form an approximate shoulder. The ZRP data from Figure 2.2 are as shown in Table 2.2.

The current location of the hand is $\mathbf{u}_a$, $\mathbf{u}_t$, and $\mathbf{P}$. The current position of the spherical wrist center H is obtained as $\mathbf{H} = \mathbf{P} - h\mathbf{u}_a$. Displacement of $\mathbf{H}_0$: $(b - a + c, 0, 0)^t$ to $\mathbf{H}$ is according to Eq. (2.6),

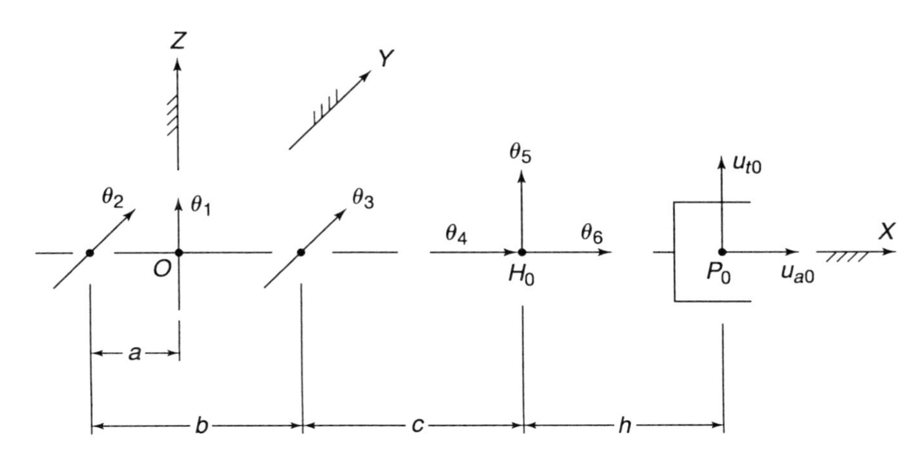

FIGURE 2.2. ZRP data for manipulator case 2.

TABLE 2.2. ZRP data for manipulator case 2.

Axis No.	Axis Type	Dir. $\mathbf{u}_{io}$	Loc. $\mathbf{Q}_{io}$
1	R	$0, 0, 1$	$0, 0, 0$
2	R	$0, 1, 0$	$-a, 0, 0$
3	R	$0, 1, 0$	$b - a, 0, 0$
4	R	$1, 0, 0$	$b - a + c, 0, 0$
5	R	$0, 0, 1$	$b - a + c, 0, 0$
6	R	$1, 0, 0$	$b - a + c, 0, 0$

Hand: $\mathbf{u}_{ao} = (1, 0, 0)$, $\mathbf{u}_{to} = (0, 0, 1)$, $\mathbf{P}_o = (b - a + c + h, 0, 0)$.

where

$$D_1 = D(\theta_1, 0, \mathbf{K}, \mathbf{O}),\ D_2 = D(\theta_2, 0, \mathbf{J}, \mathbf{Q}_{2o}),\ D_3 = D(\theta_3, 0, \mathbf{J}, \mathbf{Q}_{3o}).$$

All of these matrices can be found by inspection. Care must be exercised for D_2 and D_3 because their rotation axes do not pass through the origin, and vectors $\mathbf{d}_2$ and $\mathbf{d}_3$ are not zero. To obtain $\mathbf{d}_2$, for example, consider the XZ plane and perform a rotation by θ_2 (positive angle is Z-to-X) about $X = -a$, $Z = 0$. The body point O', which is coincident with the base origin O, will move to O'' with coordinates $X = -a(1 - \cos\theta_2)$, $Z = -a\sin\theta_2$; the Y coordinate remains unchanged. Then,

$$\mathbf{d}_2 = \mathbf{O}'\mathbf{O}'' = \{-a(1-\cos\theta_2), 0, -a\sin\theta_2\}^t$$

and

$$D_2 = \begin{bmatrix} \cos\theta_2 & 0 & \sin\theta_2 & -a(1-\cos\theta_2) \\ 0 & 1 & 0 & 0 \\ -\sin\theta_2 & 0 & \cos\theta_2 & -a\sin\theta_2 \\ 0 & 0 & 0 & 1 \end{bmatrix}.$$

After expanding Eq. (2.6), we find

$$\begin{aligned} H_x &= \cos\theta_1[c\cos(\theta_2+\theta_3) + b\cos\theta_2 - a] \\ H_y &= \sin\theta_1[c\cos(\theta_2+\theta_3) + b\cos\theta_2 - a] \\ H_z &= -c\sin(\theta_2+\theta_3) - b\sin\theta_2. \end{aligned}$$

Note that the expressions for H_x and H_y contain a common factor on the right-hand side. There are two possible choices for θ_1:

$$\theta_1 = \arctan(H_y/H_x);\quad \text{also,}\ (\pi + \theta_1).$$

Define $H_Q = H_x\cos\theta_1 + H_y\sin\theta_1 + a$. Then,

$$H_Q = c\cos(\theta_2+\theta_3) + b\cos\theta_2.$$

After squaring and adding the expressions for H_z and H_Q, we can find two choices for θ_3,

$$\theta_3 = \arccos\left(\frac{H_Q^2 + H_z^2 - b^2 - c^2}{2bc}\right),\quad \text{also } (-\theta_3).$$

Equations for H_z and H_Q are now rearranged as follows to solve for θ_2,

$$\begin{bmatrix} c\sin\theta_3 & b+c\cos\theta_3 \\ b+c\cos\theta_3 & -c\sin\theta_3 \end{bmatrix} \begin{bmatrix} \cos\theta_2 \\ \sin\theta_2 \end{bmatrix} = \begin{bmatrix} -H_z \\ H_Q \end{bmatrix}.$$

After solving for $\cos\theta_2$ and $\sin\theta_2$, we use Eq. (2.13) to find θ_2,

$$\theta_2 = 2\arctan\left(\frac{-cH_Q\sin\theta_3 - (b+c\cos\theta_3)H_z}{H_Q^2 + H_z^2 + (b+c\cos\theta_3)H_Q - cH_z\sin\theta_3}\right).$$

The wrist solution is similar to that presented for the Stanford Arm, except that the values of $\mathbf{v}_a$ [Eq. (2.10)] and $\mathbf{w}_t$ [Eq. (2.12)] must be found by using the following rotation matrices:

$$R_1 = R(\theta_1, \mathbf{K}), R_2 = R(\theta_2, \mathbf{J}), R_3 = R(\theta_3, \mathbf{J}),$$

$$R_4 = R(\theta_4, \mathbf{I}), R_5 = R(\theta_5, \mathbf{K}).$$

There are two choices for the wrist solution. Thus, overall, there are eight solution sets for this manipulator. The solution sets coalesce (i.e., singularities occur) when

(i) The wrist center H lies on the base Z axis, i.e., $H_x = H_y = 0$. Joint angle θ_1 then becomes indeterminate.

(ii) Joint angle $\theta_3 = 0$ (elbow stretched out) or π (elbow folded over).

(iii) Joint angle $\theta_5 = 0$ or π. The three wrist axes then become coplanar.

The translations of the wrist center H become restricted in cases (i) and (ii), while the orientations of the hand become restricted in case (iii).

Example 2.2. Let $a = 1'$, $b = 5'$, $c = 5'$, and $h = 1'$ in Figure 2.2. The hand is to move from its ZRP position to the current position $\mathbf{P}$: $(4', 1', 5')^t$, $\mathbf{u_a}$: $(0, 1, 0)^t$, $\mathbf{u_t}$: $(1, 0, 0)^t$. By using the closed-form solution for joint angles, find all of the eight solution sets that correspond to the current hand position.

Solution: Joint angles are found in the following sequence: θ_1 (two choices), θ_3 (two choices), θ_2, θ_5 (two choices), θ_4, θ_6. The eight solution sets are as shown in Table 2.3.

TABLE 2.3. Eight joint solutions for manipulator case 2.

Soln. #	θ_1	θ_2	θ_3	θ_4	θ_5	θ_6
1	0°	0°	−90°	0°	90°	180°
2	0°	0°	−90°	180°	−90°	0°
3	0°	−90°	90°	0°	90°	90°
4	0°	−90°	90°	180°	−90°	−90°
5	180°	−175.3°	108.7°	180°	90°	−156.6°
6	180°	−175.3°	108.7°	0°	−90°	23.36°
7	180°	−66.6°	−108.7°	180°	90°	94.7°
8	180°	−66.6°	−108.7°	0°	−90°	−85.3°

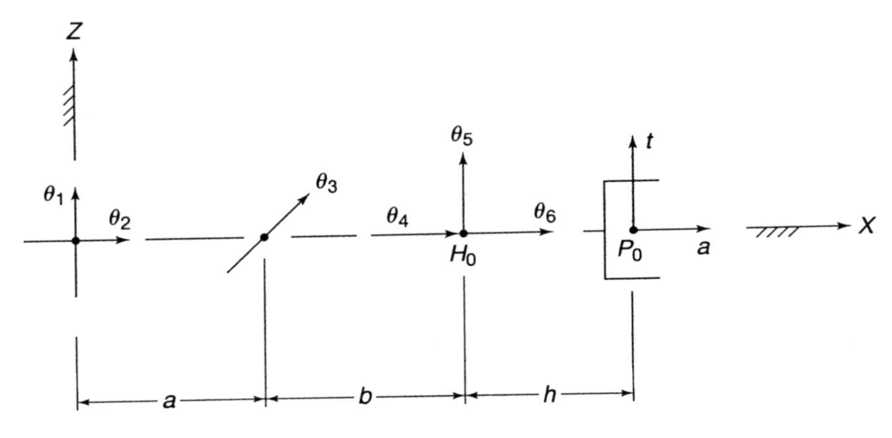

FIGURE 2.3. ZRP data for manipulator case 3.

2.5 Manipulator Case 3

The regional structure of this manipulator is $R \perp R \perp R$, but the second revolute axis is directed along the common perpendicular between the nonintersecting first and third revolute axes (Figure 2.3). The wrist has cointersecting $R \perp R \perp R$ axes. Joint 3 forms an elbow. The ZRP data from Figure 2.3 are as shown in Table 2.4.

The problem is: Given the current hand location as $\mathbf{u}_a$, $\mathbf{u}_t$ and $\mathbf{P}$, find all of the joint variables. Coordinates of point H_o are $(a+b, 0, 0)$. Again, $\mathbf{H} = \mathbf{P} - h\mathbf{u}_a$. For Eq. (2.6),

$$D_1 = D(\theta_1, 0, \mathbf{K}, \mathbf{O}),\ D_2 = D(\theta_2, 0, \mathbf{I}, \mathbf{O}),\ D_3 = D(\theta_3, 0, \mathbf{J}, \mathbf{Q}_{3o}).$$

TABLE 2.4. ZRP data for manipulator case 3.

Axis No.	Axis Type	Dir. $\mathbf{u}_{io}$	Loc. $\mathbf{Q}_{io}$
1	R	$0,0,1$	$0,0,0$
2	R	$1,0,0$	$0,0,0$
3	R	$0,1,0$	$a,0,0$
4	R	$1,0,0$	$0,0,0$
5	R	$0,0,1$	$a+b,0,0$
6	R	$1,0,0$	$0,0,0$

Hand: $\mathbf{u}_{ao} = (1,0,0)$, $\mathbf{u}_{to} = (0,0,1)$, $\mathbf{P}_o = (a+b+h,0,0)$.

All of these matrices can be written by inspection, but special attention should be paid to D_3,

$$D_3 = \begin{bmatrix} \cos\theta_3 & 0 & \sin\theta_3 & a(1-\cos\theta_3) \\ 0 & 1 & 0 & 0 \\ -\sin\theta_3 & 0 & \cos\theta_3 & a\sin\theta_3 \\ 0 & 0 & 0 & 1 \end{bmatrix}.$$

Expanding Eq. (2.6),

$$\begin{aligned} H_x &= (a+b\cos\theta_3)\cos\theta_1 - b\sin\theta_1\sin\theta_2\sin\theta_3 \\ H_y &= (a+b\cos\theta_3)\sin\theta_1 + b\cos\theta_1\sin\theta_2\sin\theta_3 \\ H_z &= -b\cos\theta_2\sin\theta_3. \end{aligned}$$

After squaring and adding the three components of $\mathbf{H}$, we can solve for the two choices of θ_3:

$$\theta_3 = \arccos\left(\frac{H_x^2 + H_y^2 + H_z^2 - a^2 - b^2}{2ab}\right), \quad \text{also } (-\theta_3).$$

Note that H_x and H_y contain a factor $(a+b\cos\theta_3)$. Useful expressions can be obtained by retaining and by eliminating this factor. For the first case, eliminate θ_2 from H_x and H_y by forming

$$H_x\cos\theta_1 + H_y\sin\theta_1 = a + b\cos\theta_3.$$

The only unknown in this equation is θ_1. Using the trick in Eq. (2.14), we can find two choices for θ_1:

$$\theta_1 = 2\arctan\left(\frac{H_y + H_Q}{H_x + a + b\cos\theta_3}\right),$$

$$H_Q = \pm\sqrt{H_x^2 + H_y^2 - (a + b\cos\theta_3)^2}.$$

Now eliminate $(a + b\cos\theta_3)$ from H_x and H_y by forming

$$-H_x \sin\theta_1 + H_y \cos\theta_1 = b\sin\theta_2 \sin\theta_3.$$

This equation gives $\sin\theta_2$, and the expression for H_z gives $\cos\theta_2$. Then, by using Eq. (2.13), we find θ_2:

$$\theta_2 = 2\arctan\left(\frac{H_y\cos\theta_1 - H_x\sin\theta_1}{b\sin\theta_3 - H_z}\right).$$

The wrist solution is similar to that presented for the Stanford Arm, except that the values of $\mathbf{v}_a$ [Eq. (2.10)] and $\mathbf{w}_t$ [Eq. (2.12)] must be found by using the following rotation matrices:

$$R_1 = R(\theta_1, \mathbf{K}), R_2 = R(\theta_2, \mathbf{I}), R_3 = R(\theta_3, \mathbf{J}),$$

$$R_4 = R(\theta_4, \mathbf{I}), R_5 = R(\theta_5, \mathbf{K}).$$

There are eight solutions sets for this manipulator. The solution sets coalesce when:

(i) Joint angle $\theta_3 = 0$ or π (elbow stretched out or folded over).

(ii) The wrist center H lies on the surface of a right circular torus of toroidal radius a and cross sectional radius b. This is a difficult physical meaning for $H_Q = 0$.

(iii) Joint angle $\theta_5 = 0$ or π. The three wrist axes then become coplanar.

2.6 PUMA by ZRP Method

This manipulator has $R \perp R \parallel R$ regional structure, with joint axes 1 and 2 intersecting to form a shoulder (Figure 2.4). The second joint has an offset along its axis in order for the regional structure links to clear the base column. Joint 3 forms an elbow to provide the manipulator with bend capability. The wrist consists of cointersecting $R \perp R \perp R$ axes. PUMA stands for Programmable Universal Machine for Assembly, and this robotic system had all electric drives, and a reasonably sophisticated controller. It was initially built by

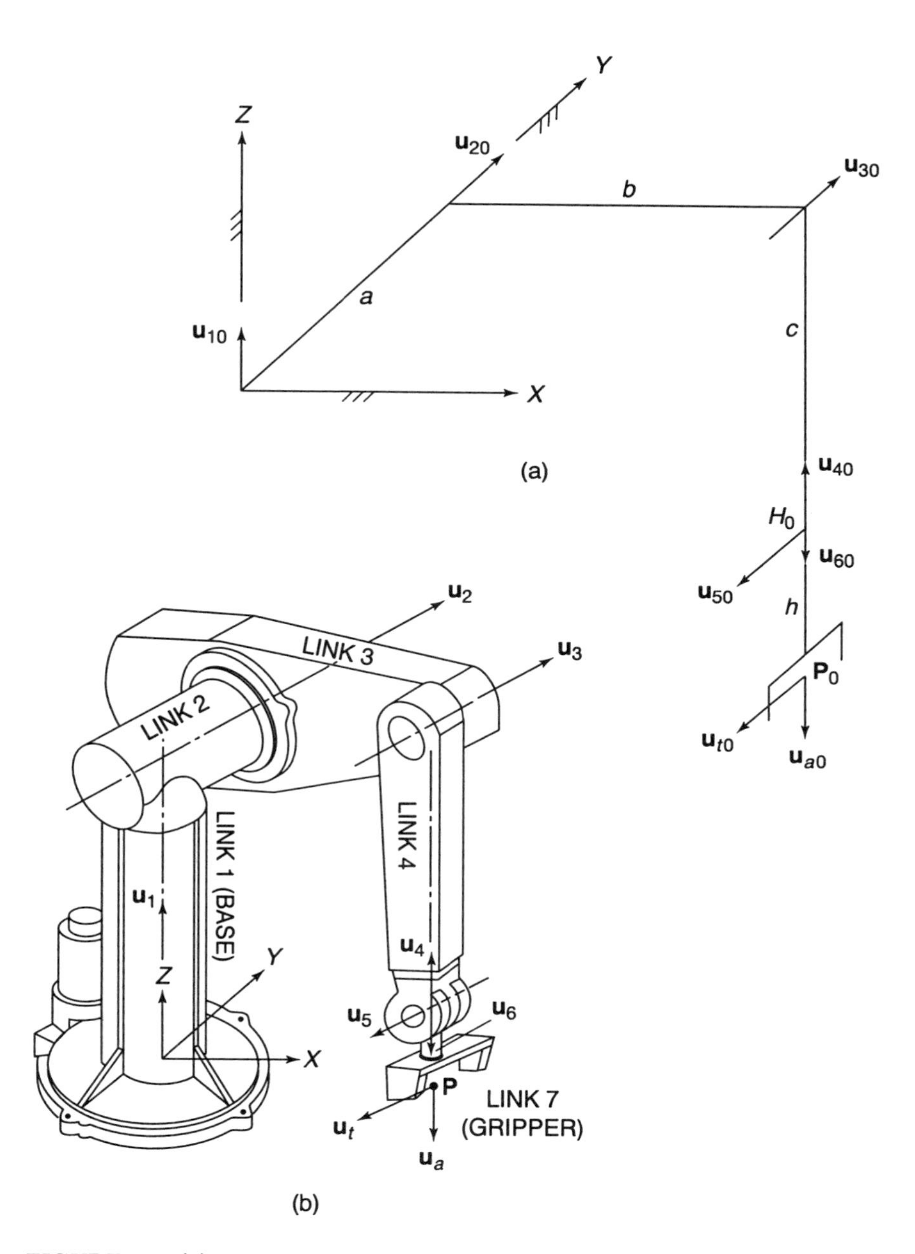

FIGURE 2.4. (a) ZRP data for the PUMA manipulator. (b) Pictorial representation of the PUMA manipulator with unit vectors $\mathbf{u}$'s marked along the joint axes.

TABLE 2.5. ZRP data for PUMA robot.

Axis No.	Axis Type	Dir. $\mathbf{u}_{io}$	Loc. $\mathbf{Q}_{io}$
1	R	$0, 0, 1$	$0, 0, 0$
2	R	$0, 1, 0$	$0, 0, 0$
3	R	$0, 1, 0$	$b, 0, 0$
4	R	$0, 0, 1$	$b, a, , 0$
5	R	$0, -1, 0$	$b, a, -c$
6	R	$0, 0, -1$	$b, a, 0$

Hand: $\mathbf{u}_{ao} = (0, 0, -1)$, $\mathbf{u}_{to} = (0, -1, 0)$, $\mathbf{P}_o = (b, a, -c - h)$.

Unimation Inc. (now defunct) to specifications developed by General Motors. Its controller could be disconnected and easily replaced by another custom-built controller. For this reason, PUMA became very popular with robotics researchers around the world. Its ZRP specifications from Figure 2.4(a) are as shown in Table 2.5.

This special zero reference position has been chosen because it is also the zero position of the DH-representation used in the Pieper–Roth method. In the next section, we will show that the results of this section (ZRP method) will be identical to those obtained by the Pieper–Roth method.

The current position of the spherical wrist center H is obtained as $\mathbf{H} = \mathbf{P} - h\mathbf{u}_a$. Displacement of $\mathbf{H}_0$: $(b, a, -c)^t$ to $\mathbf{H}$ is according to Eq. (2.6), where

$$D_1 = D(\theta_1, 0, \mathbf{K}, \mathbf{O}), \; D_2 = D(\theta_2, 0, \mathbf{J}, \mathbf{O}), \; D_3 = D(\theta_3, 0, \mathbf{J}, \mathbf{Q}_{3o}).$$

These matrices can be written by inspection. After expanding Eq. (2.6), we find

$$\begin{aligned} H_x &= \cos\theta_1\{-c\sin(\theta_2+\theta_3) + b\cos\theta_2\} - a\sin\theta_1 \\ H_y &= \sin\theta_1\{-c\sin(\theta_2+\theta_3) + b\cos\theta_2\} + a\cos\theta_1 \\ H_z &= -c\cos(\theta_2+\theta_3) - b\sin\theta_2. \end{aligned}$$

Useful expressions can be obtained by eliminating and by retaining the term $\{-c\sin(\theta_2+\theta_3) + b\cos\theta_2\}$ from H_x and H_y. Form

$$H_y\cos\theta_1 - H_x\sin\theta_1 = a$$

and solve for θ_1 by using Eq. (2.14):

$$\theta_1 = 2\arctan\left(\frac{-H_x + H_Q}{a + H_y}\right), \quad H_Q = \pm\sqrt{H_x^2 + H_y^2 - a^2}.$$

The values of θ_1 coalesce when $H_Q = 0$, i.e., when the wrist center H lies on a vertical line through the intersection point of joint axes 2 and 3.

Square and add the three components of $\mathbf{H}$, and solve for θ_3 as follows:

$$\theta_3 = \arcsin\left(\frac{a^2 + b^2 + c^2 - H_x^2 - H_y^2 - H_z^2}{2bc}\right), \quad \text{also } (\pi - \theta_3).$$

The values of θ_3 coalesce when $\theta_3 = \pm\pi/2$ (elbow stretched out or folded over).

Substituting for H_x and H_y into the definition of H_Q, we get the aforementioned factor within $\{\ldots\}$.

$$H_Q = -c\sin(\theta_2 + \theta_3) + b\cos\theta_2.$$

After rearranging the expression for H_Q and H_z as

$$\begin{bmatrix} b - c\sin\theta_3 & -c\cos\theta_3 \\ -c\cos\theta_3 & -(b - c\sin\theta_3) \end{bmatrix} \begin{bmatrix} \cos\theta_2 \\ \sin\theta_2 \end{bmatrix} = \begin{bmatrix} H_Q \\ H_z \end{bmatrix},$$

we can solve for $\cos\theta_2$ and $\sin\theta_2$ and then use Eq. (2.13) to solve for θ_2:

$$\theta_2 = 2\arctan\left(\frac{H_z(b - c\sin\theta_3) + cH_Q\cos\theta_3}{cH_z\cos\theta_3 - H_Q(b - c\sin\theta_3) + 2bc\sin\theta_3 - b^2 - c^2}\right).$$

Because of different ZRP data for the wrist, the previous wrist solution cannot be used. In Eq. (2.10),

$$R_1 = R(\theta_1, \mathbf{K}), R_2 = R(\theta_2, \mathbf{J}), R_3 = R(\theta_3, \mathbf{J}),$$

$$R_4 = R(\theta_4, \mathbf{K}), R_5 = R(\theta_5, -\mathbf{J}),$$

$$v_{ax} = \cos\theta_4\sin\theta_5, \; v_{ay} = \sin\theta_4\sin\theta_5, \; v_{az} = -\cos\theta_5.$$

The arccos function gives two possible choices for θ_5, and then the value of θ_4 is found by using Eq. (2.13):

$$\theta_5 = \arccos(-v_{az}), \text{ also } (-\theta_5); \quad \theta_4 = 2\arctan\left(\frac{v_{ay}}{\sin\theta_5 + v_{ax}}\right).$$

The two solutions for θ_5 coalesce when $\theta_5 = 0$ or π, i.e., when the three wrist axes become coplanar.

Using Eq. (2.12) and $R_6 = R(\theta_6, -\mathbf{K})$, we get

$$w_{tx} = -\sin\theta_6, \quad w_{ty} = -\cos\theta_6, \quad w_{tz} = 0,$$

and we then use Eq. (2.13) to find θ_6:

$$\theta_6 = 2\arctan\left(\frac{w_{tx}}{w_{ty} - 1}\right).$$

There are $2 \times 2 \times 2 = 8$ solution sets corresponding to the indicated choices in the solution procedure.

2.7 PUMA by Pieper–Roth Method

The Denavit–Hartenberg parameters are given in Table 2.6 (see Figure 2.5).

The next step is to write all six DH matrices A_i; recall that the A_i matrix changes coordinates from DH-system $(i+1)$ to DH-system i [Eq. (1.12)].

$$A_1 = \begin{bmatrix} c\theta_1 & 0 & -s\theta_1 & 0 \\ s\theta_1 & 0 & c\theta_1 & 0 \\ 0 & -1 & 0 & 0 \\ 0 & 0 & 0 & 1 \end{bmatrix}, \quad A_2 = \begin{bmatrix} c\theta_2 & -s\theta_2 & 0 & bc\theta_2 \\ s\theta_2 & c\theta_2 & 0 & bs\theta_2 \\ 0 & 0 & 1 & a \\ 0 & 0 & 0 & 1 \end{bmatrix}$$

$$A_3 = \begin{bmatrix} c\theta_3 & 0 & s\theta_3 & 0 \\ s\theta_3 & 0 & -c\theta_3 & 0 \\ 0 & 1 & 0 & 0 \\ 0 & 0 & 0 & 1 \end{bmatrix}, \quad A_4 = \begin{bmatrix} c\theta_4 & 0 & s\theta_4 & 0 \\ s\theta_4 & 0 & -c\theta_4 & 0 \\ 0 & 1 & 0 & -c \\ 0 & 0 & 0 & 1 \end{bmatrix}$$

$$A_5 = \begin{bmatrix} c\theta_5 & 0 & s\theta_5 & 0 \\ s\theta_5 & 0 & -c\theta_5 & 0 \\ 0 & 1 & 0 & 0 \\ 0 & 0 & 0 & 1 \end{bmatrix}, \quad A_6 = \begin{bmatrix} c\theta_6 & -s\theta_6 & 0 & 0 \\ s\theta_6 & c\theta_6 & 0 & 0 \\ 0 & 0 & 1 & h \\ 0 & 0 & 0 & 1 \end{bmatrix}.$$

Given is the hand location matrix A_h [Eq. (1.30)], and we have to find all joint angles. The wrist center H has coordinates $(0, 0, -c)$ in system 4 and (H_x, H_y, H_z) in system 1, obtained by evaluating $\mathbf{H} = \mathbf{O}_1\mathbf{P} - h\mathbf{u}_a$ in base system 1. The transformation matrix $(A_1A_2A_3)$

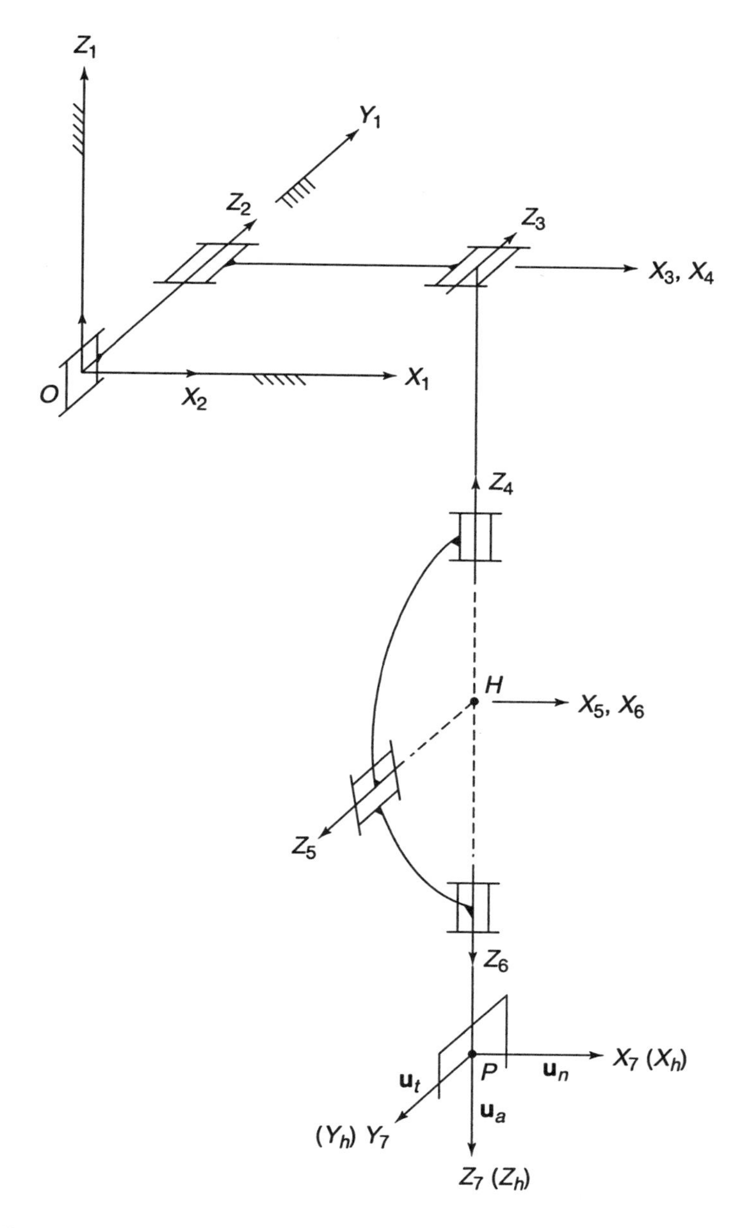

FIGURE 2.5. The X and Z directions of the DH-system for the PUMA manipulator.

TABLE 2.6. DH parameters for the PUMA robot.

i	s_i	α_i	a_i
1	0	$-90°$	0
2	a	$0°$	b
3	0	$90°$	0
4	$-c$	$90°$	0
5	0	$90°$	0
6	h	$0°$	0

changes coordinates from system 4 to system 1. After expanding this relation [see Eq. (2.2)], we find

$$\begin{aligned} H_x &= \cos\theta_1\{-c\sin(\theta_2+\theta_3)+b\cos\theta_2\} - a\sin\theta_1 \\ H_y &= \sin\theta_1\{-c\sin(\theta_2+\theta_3)+b\cos\theta_2\} + a\cos\theta_1 \\ H_z &= -c\cos(\theta_2+\theta_3) - b\sin\theta_2. \end{aligned}$$

Since these expressions are identical to those in the previous section (ZRP solution), the solutions for θ_1, θ_3, θ_2 are also the same.

Next, the axial direction of the hand (Z_h) has components $\mathbf{K}_6$: $(0, 0, 1)^t$ in system 6 and components $\mathbf{u}_a$: $(u_{ax}, u_{ay}, u_{az})^t$ in system 1. The transformation matrix to change components from system 6 to system 1 is $(R_1R_2R_3R_4R_5)$. The matrices R_i are the 3×3 principal minors of the DH matrices A_i, and note that these are different in definition from those in the previous sections.

$$\begin{aligned} R_1 &= \begin{bmatrix} c\theta_1 & 0 & -s\theta_1 \\ s\theta_1 & 0 & c\theta_1 \\ 0 & -1 & 0 \end{bmatrix}, \quad R_2 = \begin{bmatrix} c\theta_2 & -s\theta_2 & 0 \\ s\theta_2 & c\theta_2 & 0 \\ 0 & 0 & 1 \end{bmatrix}, \\ R_3 &= \begin{bmatrix} c\theta_3 & 0 & s\theta_3 \\ s\theta_3 & 0 & -c\theta_3 \\ 0 & 1 & 0 \end{bmatrix}, \quad R_4 = \begin{bmatrix} c\theta_4 & 0 & s\theta_4 \\ s\theta_4 & 0 & -c\theta_4 \\ 0 & 1 & 0 \end{bmatrix}, \\ R_5 &= \begin{bmatrix} c\theta_5 & 0 & s\theta_5 \\ s\theta_5 & 0 & -c\theta_5 \\ 0 & 1 & 0 \end{bmatrix}, \quad R_6 = \begin{bmatrix} c\theta_6 & -s\theta_6 & 0 \\ s\theta_6 & c\theta_6 & 0 \\ 0 & 0 & 1 \end{bmatrix}. \end{aligned}$$

Transforming the hand axial direction from system 6 to system 1, we get

$$\mathbf{u}_a = (R_1R_2R_3)(R_4R_5)\mathbf{K}_6.$$

Define $\mathbf{v}'_a = (R_1R_2R_3)^t\mathbf{u}_a$, and then

$$v'_{ax} = \cos\theta_4 \sin\theta_5,\ v'_{ay} = \sin\theta_4 \sin\theta_5,\ v'_{az} = -\cos\theta_5.$$

Upon expanding $(R_1R_2R_3)$, we find that the result of this product is the same for the ZRP Method and the Pieper–Roth method even though the individual R_i matrices are different. This means that $\mathbf{v}'_a$ in this section is the same as $\mathbf{v}_a$ in the previous section ($\mathbf{v}'_a = \mathbf{v}_a$), and

$$v_{ax} = \cos\theta_4 \sin\theta_5,\ v_{ay} = \sin\theta_4 \sin\theta_5,\ v_{az} = -\cos\theta_5.$$

This is now identical to the corresponding expressions obtained in the previous section, and, therefore, the solutions for θ_4 and θ_5 are also the same.

The transverse direction of the hand (Y_h) has components $\mathbf{J}_7$: $(0, 1, 0)^t$ in system 7 (hand system) and components $\mathbf{u}_t$: $(u_{tx}, u_{ty}, u_{tz})^t$ in system 1. The transformation matrix to change components from system 7 to system 1 is $(R_1R_2R_3R_4R_5R_6)$.

$$\mathbf{u}_t = (R_1R_2R_3)(R_4R_5)R_6\mathbf{J}_7.$$

Define $\mathbf{w}'_t = (R_4R_5)^t(R_1R_2R_3)^t\mathbf{u}_t$, and then

$$w'_{tx} = -\sin\theta_6,\ w'_{ty} = \cos\theta_6,\ w'_{tz} = 0.$$

When compared with the previous section, the product $(R_4R_5)^t$ above has the same first row, but the second and third rows differ by minus sign only. Then, the $\mathbf{w}'_t$ in this section and $\mathbf{w}_t$ in the previous section are related as

$$w'_{tx} = w_{tx},\ w'_{ty} = -w_{ty},\ w'_{tz} = -w_{tz}.$$

This leads to

$$w_{tx} = -\sin\theta_6,\ w_{ty} = -\cos\theta_6,\ w_{tz} = 0,$$

which is identical to that in the previous section, and the θ_6 solution is also the same.

The ZRP method solution (Section 2.6) is identical to the Pieper–Roth method solution (Section 2.7) because the zero reference position used for the ZRP method was the same as the zero position of the DH representation.

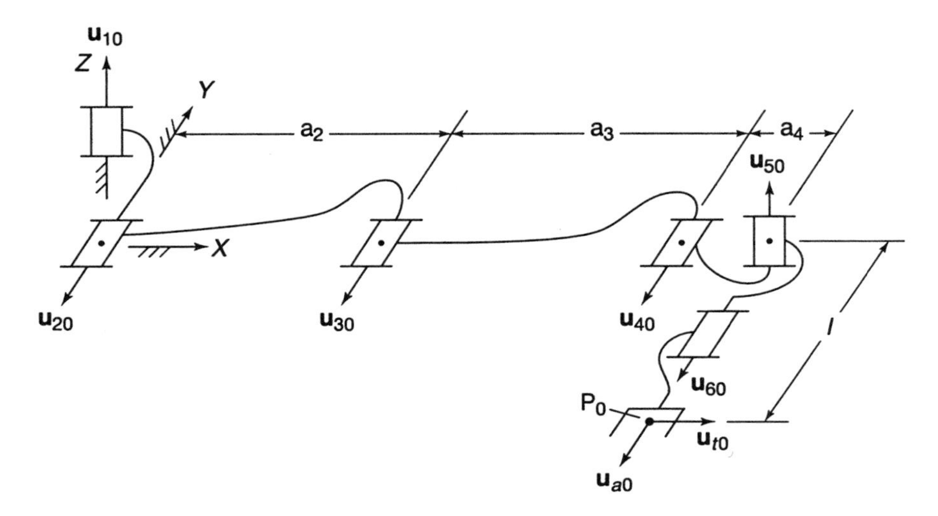

FIGURE 2.6. ZRP data for the Cincinnati Milacron T3 manipulator.

2.8 Cincinnati Milacron T3

This manipulator has six revolute axes arranged as $R \perp R \parallel R \parallel R \perp R \perp R$. It was part of an early electric–hydraulic powered robotic system. It had a shoulder ($R \perp R$), an elbow (third joint), and a wrist whose three axes did not intersect at a common point (Figure 2.6). Point H, which was critical in previous analyses, cannot be defined here. However, simplification in analysis occurs because three successive axes in the middle (axes 2, 3, 4) are parallel. The ZRP data are shown in Table 2.7.

For convenience in writing, define

$$c_i = \cos\theta_i, \quad s_i = \sin\theta_i$$

$$c_{23} = \cos(\theta_2 + \theta_3), \quad s_{23} = \sin(\theta_2 + \theta_3)$$

$$c_{234} = \cos(\theta_2 + \theta_3 + \theta_4), \quad s_{234} = \sin(\theta_2 + \theta_3 + \theta_4).$$

The joint displacement matrices $D_i = D(\theta_i, 0, \mathbf{u}_{io}, \mathbf{Q}_{io})$ are as follows:

$$D_1 = \begin{bmatrix} c_1 & -s_1 & 0 & 0 \\ s_1 & c_1 & 0 & 0 \\ 0 & 0 & 1 & 0 \\ 0 & 0 & 0 & 1 \end{bmatrix}, \quad D_2 = \begin{bmatrix} c_2 & 0 & -s_2 & 0 \\ 0 & 1 & 0 & 0 \\ s_2 & 0 & c_2 & 0 \\ 0 & 0 & 0 & 1 \end{bmatrix},$$

TABLE 2.7. ZRP data for the Cincinnati Milacron T^3.

Axis No.	Axis Type	Dir. $\mathbf{u}_{io}$	Loc. $\mathbf{Q}_{io}$
1	R	$0,0,1$	$0,0,0$
2	R	$0,-1,0$	$0,0,0$
3	R	$0,-1,0$	$a_2,0,0$
4	R	$0,-1,0$	$a_2+a_3,0,0$
5	R	$0,0,1$	$a_2+a_3+a_4,0,0$
6	R	$0,-1,0$	$a_2+a_3+a_4,0,0$

Hand: $\mathbf{u}_{ao} = (0,-1,0)$, $\mathbf{u}_{to} = (1,0,0)$, $\mathbf{P}_o = (a_2+a_3+a_4,-l,0)$.

$$D_3 = \begin{bmatrix} c_3 & 0 & -s_3 & a_2(1-c_3) \\ 0 & 1 & 0 & 0 \\ s_3 & 0 & c_3 & -a_2 s_3 \\ 0 & 0 & 0 & 1 \end{bmatrix},$$

$$D_4 = \begin{bmatrix} c_4 & 0 & -s_4 & (a_2+a_3)(1-c_4) \\ 0 & 1 & 0 & 0 \\ s_4 & 0 & c_4 & -(a_2+a_3)s_4 \\ 0 & 0 & 0 & 1 \end{bmatrix},$$

$$D_5 = \begin{bmatrix} c_5 & -s_5 & 0 & (a_2+a_3+a_4)(1-c_5) \\ s_5 & c_5 & 0 & -(a_2+a_3+a_4)s_5 \\ 0 & 0 & 1 & 0 \\ 0 & 0 & 0 & 1 \end{bmatrix},$$

$$D_6 = \begin{bmatrix} c_6 & 0 & -s_6 & (a_2+a_3+a_4)(1-c_6) \\ 0 & 1 & 0 & 0 \\ s_6 & 0 & c_6 & -(a_2+a_3+a_4)s_6 \\ 0 & 0 & 0 & 1 \end{bmatrix}.$$

The displacement of point P from P_o can be written as

$$\begin{bmatrix} \mathbf{P} \\ 1 \end{bmatrix} = D_1 D_2 D_3 D_4 D_5 \begin{bmatrix} \mathbf{P}_o \\ 1 \end{bmatrix}. \tag{a}$$

Note that D_6 does not appear in this equation because it does not affect point P, which lies on the axis of the sixth revolute joint. Axes 2, 3, 4 are parallel, and we will keep matrices $(D_2D_3D_4)$ together. There are too many matrices on the right-hand side, and we would like to shift some of them to the left-hand side. But the only matrix that can be moved in this way is D_1; then,

$$D_1^{-1}\begin{bmatrix}\mathbf{P}\\1\end{bmatrix} = D_2D_3D_4D_5\begin{bmatrix}\mathbf{P}_o\\1\end{bmatrix}. \tag{b}$$

The rotational part (3×3 principal minor) will give a corresponding expression for the movement of $\mathbf{u}_a$ from $\mathbf{u}_{ao}$.

$$R_1^t\mathbf{u}_a = R_2R_3R_4R_5\mathbf{u}_{ao}. \tag{c}$$

Equations (b) and (c) are now expanded. Coordinates of point P are (X, Y, Z).

$$\begin{aligned} Xc_1 + Ys_1 &= (a_4 + ls_5)c_{234} + a_3c_{23} + a_2c_2 \\ -Xs_1 + Yc_1 &= -lc_5 \\ Z &= (a_4 + ls_5)s_{234} + a_3s_{23} + a_2s_2 \end{aligned} \tag{d}$$

$$\begin{aligned} u_{ax}c_1 + u_{ay}s_1 &= s_5c_{234} \\ -u_{ax}s_1 + u_{ay}c_1 &= -c_5 \\ u_{az} &= s_5s_{234}. \end{aligned} \tag{e}$$

The strategy will be to find θ_1 and θ_5 from the second and fourth equations, $(\theta_2 + \theta_3 + \theta_4)$ from the fourth and sixth equations, and θ_2 and $(\theta_2 + \theta_3)$ from the first and third equations. From the second equations of sets (d) and (e), eliminate c_5 to obtain two choices for θ_1.

$$\theta_1 = \arctan\left(\frac{Y - lu_{ay}}{X - lu_{ax}}\right), \quad \text{also } (\pi + \theta_1).$$

The joint angle becomes indeterminate when the argument becomes $(0/0)$, and that occurs when the center of joint 5 lies on the first joint axis.

Two choices for joint angle θ_5 can be found from the second equation of set (e),

$$\theta_5 = \arccos(u_{ax}s_1 - u_{ay}c_1), \quad \text{also } (-\theta_5).$$

The θ_5 values coalesce when $\theta_5 = 0$ or π, i.e., when joint axis 6 also becomes parallel to joint axis group 2, 3, 4 (see Figure 2.6).

Now the first and third equations of set (e) and Eq. (2.13) allow us to solve for the sum $(\theta_2 + \theta_3 + \theta_4)$:

$$\theta_2 + \theta_3 + \theta_4 = 2\arctan\left(\frac{u_{az}}{s_5 + c_1u_{ax} + s_1u_{ay}}\right).$$

In the first and third equations of set (d), the unknowns at this stage are θ_2 and the sum $(\theta_2 + \theta_3)$. Eliminating $(\theta_2 + \theta_3)$ by using $(c_{23})^2 + (s_{23})^2 = 1$, we get an equation in θ_2 of the form

$$A \sin\theta_2 + B\cos\theta_2 + C = 0.$$

This can be solved for two choices for θ_2 as shown in Eq. (2.14).

$$\theta_2 = 2\arctan\left(\frac{-A \pm \sqrt{A^2 + B^2 - C^2}}{C - B}\right),$$

where

$$\begin{aligned} A &= 2a_2 g,\ B = 2a_2 f,\ C = a_2^2 + f^2 + g^2 - a_3^2, \\ f &= (a_4 + l s_5)c_{234} - X c_1 - Y s_1,\ g = (a_4 + l s_5)s_{234} - Z. \end{aligned}$$

The values coalesce when the parallel joint axis group 2, 3, 4 also becomes coplanar (see Figure 2.6), i.e., the elbow is stretched out or folded over ($\theta_3 = 0$ or π). This observation is deduced from physical insight rather than from a formal mathematical investigation of the above solution, but it can be verified by means of numerical examples.

The eliminated variable group $(\theta_2 + \theta_3)$ can now be solved from c_{23}, s_{23}, and Eq. (2.13).

$$\theta_2 + \theta_3 = 2\arctan\left(\frac{g + a_2 s_2}{f + a_2 c_2 - a_3}\right).$$

Because θ_2, $(\theta_2+\theta_3)$, and $(\theta_2+\theta_3+\theta_4)$ are now known, the joint angles θ_3 and θ_4 are also known. The only variable left to be determined is θ_6. The expression for the movement of $\mathbf{u}_t$ from $\mathbf{u}_{to}$ is as follows (note the presence of matrix R_6):

$$R_1^t \mathbf{u}_t = R_2 R_3 R_4 R_5 (R_6 \mathbf{u}_{to}). \tag{f}$$

After expansion,

$$\begin{aligned} u_{tx}c_1 + u_{ty}s_1 &= c_{234}c_5c_6 - s_{234}s_6 \\ -u_{tx}s_1 + u_{ty}c_1 &= s_5c_6 \\ u_{tz} &= s_{234}c_5c_6 + c_{234}s_6. \end{aligned} \tag{g}$$

The first and third equations of set (g) can be solved as linear equations for c_6 and s_6 and from Eq. (2.13) θ_6 can be found as follows:

$$\theta_6 = 2 \arctan \left(\frac{c_{234} c_5 u_{tz} - s_{234} c_5 (c_1 u_{tx} + s_1 u_{ty})}{c_5 + c_{234}(c_1 u_{tx} + s_1 u_{ty}) + s_{234} u_{tz}} \right).$$

This manipulator has $2 \times 2 \times 2 = 8$ solution sets. The reader may be curious about the unusual zero reference position that was chosen for this manipulator. However, there are two reasons why this position was chosen. First, in this position, we can see two of the singularities ($\theta_3 = 0$ or π, and $\theta_5 = 0$ or π). Second, if the Denavit–Hartenberg coordinate systems are shown, and the Pieper–Roth method is used, then this position is also the zero position of the DH representation (in this case, the X axis of the hand system must point in the transverse direction, and that will require a minor adjustment (exchange of columns 1 and 2) in the definitions of the hand location matrices A_h and A_{ho} in Eq. (1.32). All of the numerical results for this manipulator, which are obtained by using the above expressions derived from the ZRP method, can then be compared directly with those obtained from the Pieper–Roth method.

2.9 Geared Wrists

The three-revolute, serial wrist with joint mounted motors is quite bulky. Because of the large wrist mass far away from the base column, this results in lower payload capacity, poor dynamics, and undesirable vibrations. A design that is both compact and sturdy is based upon a system of bevel gears (Figure 2.7). It is an ingenious modification of the differential mechanism. The basic skeleton in Figure 2.7(a) is formed by bodies B_5—left fork, B_6—right fork, and B_7—gripper. The wrist degrees-of-freedom are θ_4 (between B_4 and B_5), θ_5 (between B_5 and B_6), θ_6 (between B_6 and B_7). These degrees-of-freedom are powered through bevel-geared connections: $G_1 - G_3 - G_4$ and $G_2 - G_5 - G_4$. A characteristic feature is the oversized bevel gear G_4 along joint axis 6. All bevel gears have a common apex at point H (wrist center). The actuation is remote from the wrist and is through a set of coaxial tubes connected to gears G_1 and G_2 and body B_5. The relation among the actuation variables ϕ_1, ϕ_2, ϕ_3 (at the coaxial tubes) and wrist joint variables θ_4, θ_5, θ_6 can be derived as follows.

If we start the analysis from given actuation variables ϕ_1, ϕ_2, ϕ_3,

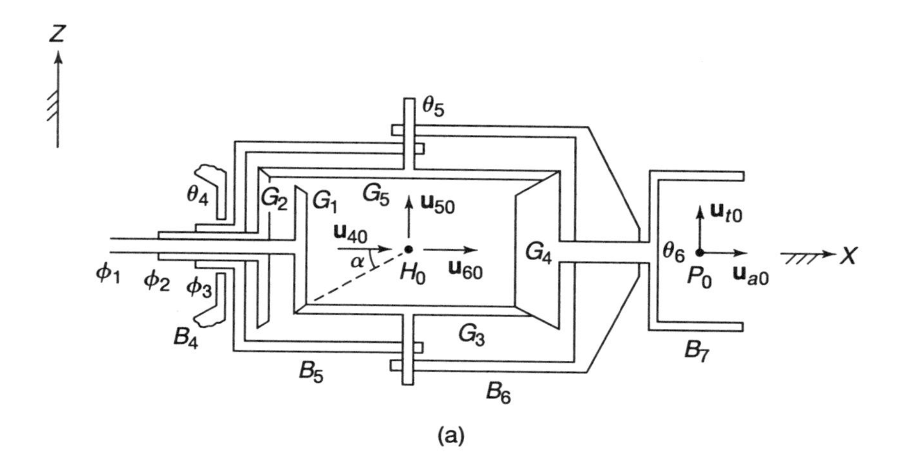

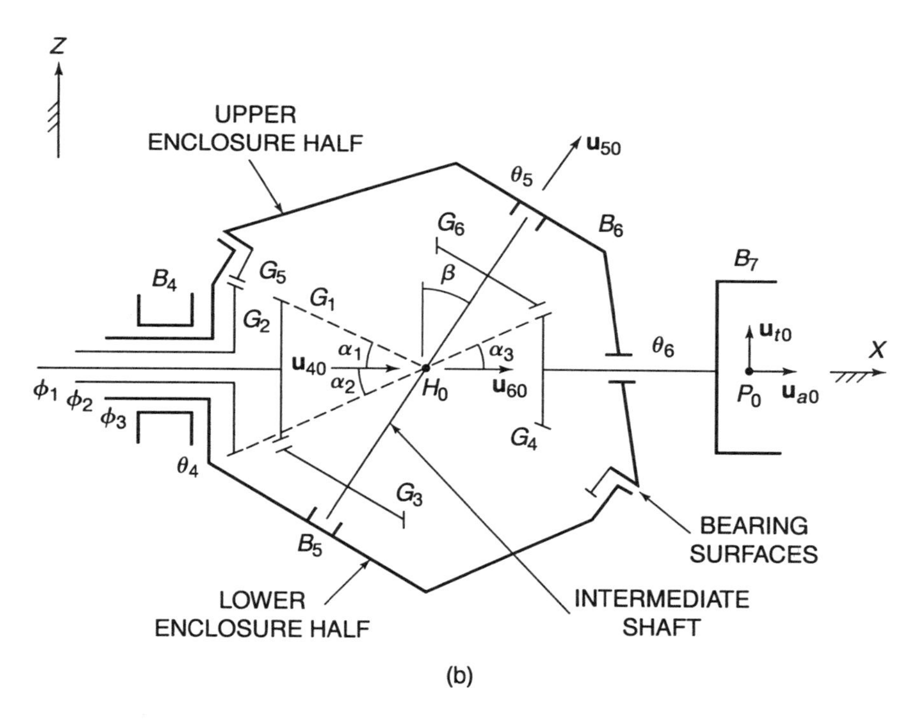

FIGURE 2.7. (a) Kinematic structure of an orthogonal, bevel-geared, three-roll wrist. (b) Kinematic structure of an oblique, bevel-geared wrist with continuous (or complete) three-roll.

and attempt to find wrist joint variables θ_4, θ_5, θ_6, then we encounter a spatial planetary motion, which is difficult to analyze. On the other hand, if we start from the given wrist joint variables θ_4, θ_5, θ_6, then we can find the actuation variables ϕ_1, ϕ_2, ϕ_3 more simply by using the principle of linear superposition. We analyze three simple gear trains for the effects of changing the wrist joint variables one at a time from the zero reference position (ZRP) of the wrist, which is defined as

$$\begin{aligned} \phi_1 &= \phi_2 = \phi_3 = 0, \quad \theta_4 = \theta_5 = \theta_6 = 0 \\ \mathbf{u}_{40} &= \mathbf{u}_{60} = \mathbf{u}_{ao} = \mathbf{I}, \quad \mathbf{u}_{50} = \mathbf{u}_{to} = \mathbf{J}. \end{aligned}$$

Let $t = \tan\alpha$, where α is half of the pitch cone angle of the gears on actuation shafts 1 and 2, and joint axis 6. The bevel gears in mesh must have their apexes at a common point ($\mathbf{H}_o$ in this case); therefore, the half of the pitch cone angle of the gears along joint axis 5 is $[(\pi/2) - \alpha]$.

The first change that is made is $(\theta_4, \theta_5, \theta_6) = (x, 0, 0)$, i.e., the fourth wrist joint angle is changed by amount x, but the fifth and sixth wrist joints are locked. The entire wrist then rotates as a rigid body about the horizontal axis in Figure 2.7(a), and all three actuation angles (ϕ's) are changed by amount x (Table 2.8). The second change, made with respect to ZRP, is $(\theta_4, \theta_5, \theta_6) = (0, y, 0)$, i.e., the fourth and sixth wrist joints are frozen, but the fifth wrist joint angle is changed by amount y. As the right fork (body B_6) rotates relative to the left fork (body B_5), with large bevel gear G_4 frozen in its bearing, gears G_3 and G_5 are rotated by amount y about their axes. This in turn causes rotation in gears G_1 and G_2 by amounts shown in Table 2.8. The third change, also made with respect to ZRP, is $(\theta_4, \theta_5, \theta_6) = (0, 0, z)$, i.e., the fourth and fifth wrist joints are frozen, and the sixth wrist joint angle is changed by amount z. Gear G_4 is rotated by amount z about its axis (joint 6), and its motion is transmitted through "idler" gear G_3 to gear G_1, and through "idler" gear G_5 to gear G_2, by amounts shown in Table 2.8. The "sum" row adds up these changes.

With $\theta_4 = x$, $\theta_5 = y$, $\theta_6 = z$, we obtain the following relations from the "sum" row:

$$\begin{bmatrix} \phi_1 \\ \phi_2 \\ \phi_3 \end{bmatrix} = \begin{bmatrix} 1 & -1/t & -1 \\ 1 & 1/t & -1 \\ 1 & 0 & 0 \end{bmatrix} \begin{bmatrix} \theta_4 \\ \theta_5 \\ \theta_6 \end{bmatrix}, \tag{2.15}$$

TABLE 2.8. Superposition table for orthogonal bevel-geared wrist.

	θ_4	θ_5	θ_6	ϕ_1	ϕ_2	ϕ_3
	x	0	0	x	x	x
	0	y	0	$-y/t$	y/t	0
	0	0	z	$-z$	$-z$	0
Sum	x	y	z	$x-y/t-z$	$x+y/t-z$	x

and the inverse relations are

$$\begin{bmatrix}\theta_4\\ \theta_5\\ \theta_6\end{bmatrix} = \begin{bmatrix}0 & 0 & 1\\ -t/2 & t/2 & 0\\ -1/2 & -1/2 & 1\end{bmatrix}\begin{bmatrix}\phi_1\\ \phi_2\\ \phi_3\end{bmatrix}. \tag{2.16}$$

In the bevel-geared wrist discussed above, joint angles θ_4 and θ_6 are unrestricted, but joint angle θ_5 is restricted because of mechanical interference. A design that has unrestricted rotations at joints 4, 5, and 6 is called an oblique bevel-geared wrist [Figure 2.7(b)]. In this wrist design, joint axes 4, 5, and 6 cointersect, but the angle between joint axes 4 and 5 is $(90° + \beta)$ and between axes 5 and 6 is $(90° - \beta)$. The characteristic features are the intermediate shaft carrying gears G_3 and G_6 and the enclosure halves formed by bodies B_5 (lower cup) and B_6 (upper cup). The enclosure halves can rotate with respect to each other, and their bearing surfaces are marked. The actuation coaxial tubes protrude on the left from the lower enclosure half, and the joint axis 6 protrudes on the right from the upper enclosure half.

The half-pitch-cone angles of the gears are presented in Table 2.9.

Starting from the given wrist joint variables θ_4, θ_5, θ_6, we find the actuation variables ϕ_1, ϕ_2, ϕ_3 by using the principle of linear superposition. We analyze three simple gear trains for the effects of changing the wrist joint variables one at a time from the zero reference position of the wrist, and the changes are made as $(\theta_4, \theta_5, \theta_6) = (x, 0, 0)$, $(0, y, 0)$, $(0, 0, z)$. The final relations are as follows:

$$\begin{aligned}\phi_1 &= \theta_4 - \frac{\cos(\alpha_1+\beta)}{\sin\alpha_1}\theta_5 + \frac{\sin\alpha_3\cos(\alpha_1+\beta)}{\sin\alpha_1\cos(\alpha_3+\beta)}\theta_6\\ \phi_2 &= \theta_4 + \frac{\cos(\alpha_2-\beta)}{\sin\alpha_2}\theta_5, \quad \phi_3 = \theta_4.\end{aligned} \tag{2.17a}$$

TABLE 2.9. α's for oblique bevel-geared wrist.

Gear	Half-Pitch-Cone Angle
G_1	α_1
G_2	α_2
G_3	$90^\circ - \beta - \alpha_1$
G_4	α_3
G_5	$90^\circ + \beta - \alpha_2$
G_6	$90^\circ - \beta - \alpha_3$

The inverse relations are as follows:

$$\theta_4 = \phi_3, \quad \theta_5 = \frac{\sin\alpha_2}{\cos(\alpha_2 - \beta)}\{\phi_2 - \phi_3\}$$

$$\theta_6 = \frac{\cos(\alpha_3 + \beta)}{\sin\alpha_3}\left\{\frac{\sin\alpha_1}{\cos(\alpha_1 + \beta)}\phi_1 + \frac{\sin\alpha_2}{\cos(\alpha_2 - \beta)}\phi_2 - \left(\frac{\sin\alpha_1}{\cos(\alpha_1 + \beta)} + \frac{\sin\alpha_2}{\cos(\alpha_2 - \beta)}\right)\phi_3\right\}. \tag{2.17b}$$

2.10 Velocity Relations

Single Rigid Body

Consider a body in the zero-reference (ZRP) and current (CP) positions [Figure 2.8(a)]. The displacement of point P from its ZRP position P_o to the current position (P) is given as

$$\begin{bmatrix} \mathbf{P} \\ 1 \end{bmatrix} = [D]\begin{bmatrix} \mathbf{P}_o \\ 1 \end{bmatrix}. \tag{2.18}$$

Differentiating this relation, we get the following relation for velocity $\mathbf{v}$ of point P:

$$\begin{bmatrix} \mathbf{v} \\ 0 \end{bmatrix} = [dD/dt]\begin{bmatrix} \mathbf{P}_o \\ 1 \end{bmatrix}. \tag{2.19}$$

Note that the fourth element in the 4×1 representation for velocity is 0 (not 1). Although correct, Eq. (2.19) is not a good expression for velocity because it relates velocity at the current position (CP) to the parameters of displacement from the ZRP to CP position and to the location of the point in ZRP position. However, we know that velocity is a local property that can be determined by observations around the current position (CP) only. Eliminating $\mathbf{P}_o$ from the velocity expression,

$$\begin{bmatrix} \mathbf{v} \\ 0 \end{bmatrix} = [(dD/dt)D^{-1}] \begin{bmatrix} \mathbf{P} \\ 1 \end{bmatrix}, \tag{2.20}$$

where

$$\begin{aligned} (dD/dt)D^{-1} &= \begin{bmatrix} dR/dt & d\mathbf{d}/dt \\ \mathbf{0} & 0 \end{bmatrix} \begin{bmatrix} R^t & -R^t\mathbf{d} \\ \mathbf{0} & 1 \end{bmatrix} \\ &= \begin{bmatrix} (dR/dt)R^t & -(dR/dt)R^t\mathbf{d} + (d\mathbf{d}/dt) \\ \mathbf{0} & 0 \end{bmatrix}. \end{aligned} \tag{2.21}$$

The right-hand side of Eq. (2.20) now contains the current position $\mathbf{P}$ of the point in question. From the local nature of velocity, we can furthermore deduce that the matrix $[(dD/dt)D^{-1}]$ must not depend upon the parameters of ZRP to CP displacement, even though these parameters are present in the expression above; i.e., in forming the product $[(dD/dt)D^{-1}]$, the effects of the ZRP position should cancel out. This can be seen by developing an alternate expression for velocity from the first principles.

Let us consider an extension of the body such that it includes a body point O' which is coincident with the base system origin O. Then, we have the following vector relation, which relates the velocities of two points (O' and P) in the rigid body:

$$\mathbf{v}_P = \mathbf{v}_{O'} + \boldsymbol{\omega} \times \mathbf{O'P}. \tag{2.22}$$

Here, $\boldsymbol{\omega}$ is the angular velocity of the body, and $\mathbf{O'P} = \mathbf{OP} = \mathbf{P}$. By changing the angular velocity vector into a 3×3 skew-symmetric form $[\Omega]$, the above vector relation can be written in the matrix form as

$$\begin{bmatrix} \mathbf{v} \\ 0 \end{bmatrix} = \begin{bmatrix} \Omega & \mathbf{v}_{O'} \\ \mathbf{0} & 0 \end{bmatrix} \begin{bmatrix} \mathbf{P} \\ 1 \end{bmatrix}, \tag{2.23}$$

where

$$[\Omega] = \begin{bmatrix} 0 & -\omega_z & \omega_y \\ \omega_z & 0 & -\omega_x \\ -\omega_y & \omega_x & 0 \end{bmatrix}.$$

Comparing the two different expressions for velocity of point P, we find the matrix $[(dD/dt)D^{-1}]$ in terms of local quantities, as was our objective:

$$(dD/dt)D^{-1} = \begin{bmatrix} \Omega & \mathbf{v}_{O'} \\ \mathbf{0} & 0 \end{bmatrix} = [V(\omega, v_{\mathrm{ISA}}, \mathbf{u}_{\mathrm{ISA}}, \mathbf{Q}_{\mathrm{ISA}})]. \tag{2.24}$$

This 4×4 matrix is defined as the velocity matrix $[V(\omega, v_{\text{ISA}}, \mathbf{u}_{\text{ISA}}, \mathbf{Q}_{\text{ISA}})]$, where the subscript ISA represents the unique instantaneous velocity axis, or the axis of instantaneous turn-slide [Figure 2.8(a)]. The first-order variations (or derivatives) of $\mathbf{u}_{\text{ISA}}$ and $\mathbf{Q}_{\text{ISA}}$ are zero, but the second- and higher-order variations are not zero. The direction of ISA is given by the unit vector $\mathbf{u}_{\text{ISA}}$, its location by the position vector of a point on the ISA $\mathbf{Q}_{\text{ISA}}$; the body has an angular speed ω about the ISA and a linear (sliding) speed v_{ISA} along the ISA. Clearly, the velocity matrix $[V]$ depends upon local properties of motion around the current position (CP). By comparing the elements of matrices in Eqs. (2.21) and (2.24),

$$[\Omega] = [(dR/dt)R^t], \quad \mathbf{v}_{o'} = -[(dR/dt)R^t]\mathbf{d} + (d\mathbf{d}/dt). \tag{2.25}$$

Note the important relation between the skew-symmetric angular velocity matrix $[\Omega]$ and the rotation matrix $[R]$; the former depends upon the local properties of motion around the current position (CP), while the latter depends upon both the ZRP and CP positions. The second relation, for the velocity of the body point coincident with the base origin, can be written in vector form as

$$\mathbf{v}_{o'} = (d\mathbf{d}/dt) - \boldsymbol{\omega} \times \mathbf{d}. \tag{2.26}$$

This is not a good expression because the vector $\mathbf{d}$ on the right-hand side does not represent a local quantity. However, in terms of the parameters of the ISA [Figure 2.8(a)], this can also be written, from the first principles, as

$$\mathbf{v}_{o'} = (v_{\text{ISA}}\mathbf{u}_{\text{ISA}}) + (\omega\mathbf{u}_{\text{ISA}}) \times (-\mathbf{Q}_{\text{ISA}}), \tag{2.27}$$

and the local nature of velocity $\mathbf{v}_{o'}$ is then obvious from the right-hand-side terms; the corresponding matrix form is

$$\mathbf{v}_{o'} = v_{\text{ISA}}\mathbf{u}_{\text{ISA}} - \omega U_{\text{ISA}}\mathbf{Q}_{\text{ISA}}, \tag{2.28}$$

where $[U_{\text{ISA}}]$ is the 3×3 skew-symmetric form of the unit vector $\mathbf{u}_{\text{ISA}}$.

Recall from (1.19):

$$[U] = \begin{bmatrix} 0 & -u_z & u_y \\ u_z & 0 & -u_x \\ -u_y & u_x & 0 \end{bmatrix}.$$

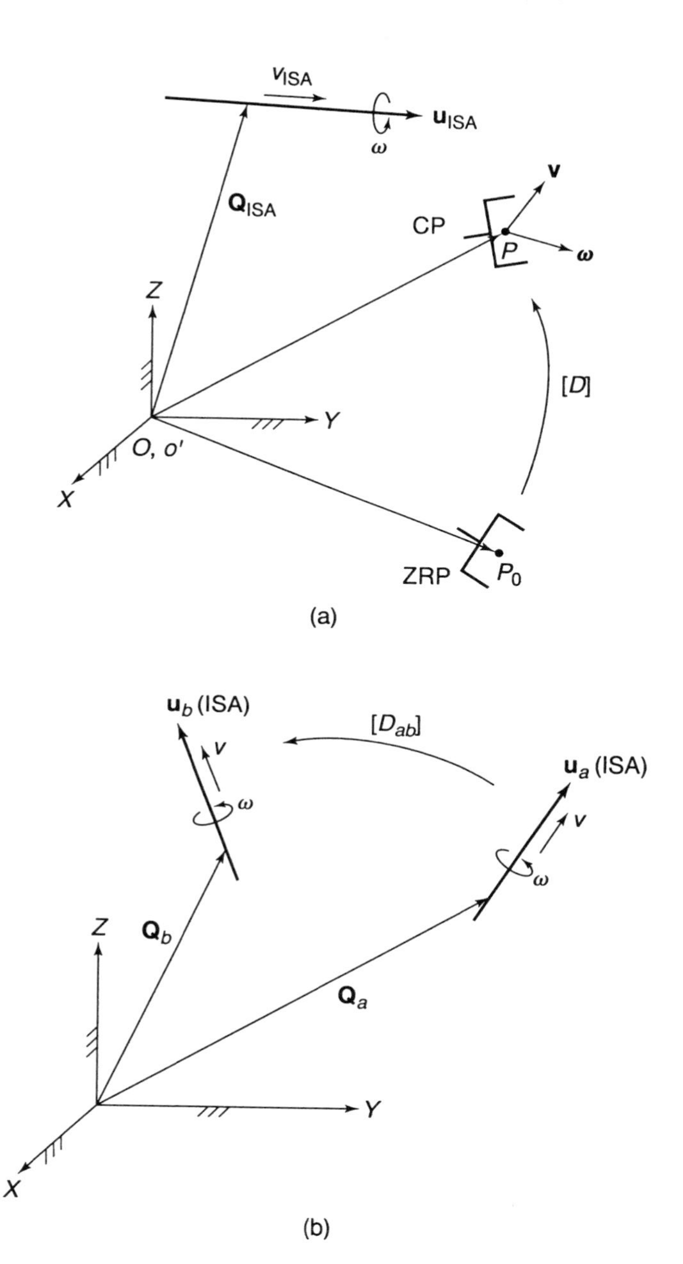

FIGURE 2.8. (a) The ZRP position, the current position (CP), the current velocity state, and the instantaneous turn-slide (ISA) representation of the current velocity state. (b) Shifted instantaneous turn-slides (ISAs) with the same (ω, v) used in the velocity similarity principle.

An alternate functional representation of the velocity matrix $[V]$—see the middle term of Eq. (2.24)—in terms of the parameters of ISA is then as follows:

$$[V(\omega, v_{\mathrm{ISA}}, \mathbf{u}_{\mathrm{ISA}}, \mathbf{Q}_{\mathrm{ISA}})] = \begin{bmatrix} \Omega & v_{\mathrm{ISA}}\mathbf{u}_{\mathrm{ISA}} - \omega U_{\mathrm{ISA}}\mathbf{Q}_{\mathrm{ISA}} \\ \mathbf{0} & 0 \end{bmatrix}. \quad (2.29)$$

The velocity matrix can be viewed as the following operator to change the current position data into the velocity data:

$$\begin{bmatrix} \mathbf{v} \\ 0 \end{bmatrix} = [V(\omega, v_{\mathrm{ISA}}, \mathbf{u}_{\mathrm{ISA}}, \mathbf{Q}_{\mathrm{ISA}})] \begin{bmatrix} \mathbf{P} \\ 1 \end{bmatrix}. \quad (2.30)$$

Serial Link Manipulator

The following result, stated without proof, will be useful in the derivations. Let a body have the instantaneous velocity axis (ISA) as $(\mathbf{u}_a, \mathbf{Q}_a)$ and the rotational speed about the ISA and the translational speed along the ISA as (ω, v). The body is then displaced [Figure 2.8(b)] such that the new ISA is $(\mathbf{u}_b, \mathbf{Q}_b)$; the corresponding displacement matrix is $[D_{ab}]$. Speed parameters for the new ISA are the same as before, i.e., (ω, v). Then, we have the following relation between the two velocity matrices:

$$[V(\omega, v, \mathbf{u}_b, \mathbf{Q}_b)] = [D_{ab}][V(\omega, v, \mathbf{u}_a, \mathbf{Q}_a)][D_{ab}]^{-1}. \quad (2.31)$$

This is called the velocity similarity principle.

Recall the definition of the ith joint displacement matrix D_i as

$$D_i = D(\theta_i, s_i, \mathbf{u}_{io}, \mathbf{Q}_{io}) = \begin{bmatrix} R_i & s_i\mathbf{u}_{io} - (R_i - I)\mathbf{Q}_{io} \\ \mathbf{0} & 1 \end{bmatrix}, \quad (2.4)$$

where

$$R_i = R(\theta_i, \mathbf{u}_{io}) = [I + U_{io}\sin\theta_i + U_{io}^2(1 - \cos\theta_i)] \quad (1.17)$$

and the ZRP data for the joint axis are $(\mathbf{u}_{io}, \mathbf{Q}_{io})$, $i = 1, 6$. When D_i is differentiated, $\mathbf{u}_{io}$, U_{io}, and $\mathbf{Q}_{io}$ must be treated as constants but $\theta_i = \theta_i(t)$ and $s_i = s_i(t)$. Then, after utilizing the properties of the skew-symmetric matrices, $U^3 = -U$, $U^4 = -U^2$, we obtain the following after some simplifications:

$$(dD_i/dt)D_i^{-1} = \begin{bmatrix} (d\theta_i/dt)U_{io} & (ds_i/dt)\mathbf{u}_{io} - (d\theta_i/dt)U_{io}\mathbf{Q}_{io} \\ \mathbf{0} & 0 \end{bmatrix}$$

$$= [V(d\theta_i/dt, ds_i/dt, \mathbf{u}_{io}, \mathbf{Q}_{io})]. \tag{2.32}$$

This is a special case of the general result in Eq. (2.24). Also recall the governing equation of the serial manipulator,

$$D_h = D_1 D_2 D_3 D_4 D_5 D_6. \tag{2.5}$$

Differentiating and postmultiplying by D_h^{-1}, we get

$$\begin{aligned}[(dD_h/dt)D_h^{-1}] = {} & [(dD_1/dt)D_1^{-1}] + [D_1][(dD_2/dt)D_2^{-1}][D_1]^{-1} \\ & + [D_1 D_2][(dD_3/dt)D_3^{-1}][D_1 D_2]^{-1} + \cdots \\ & + [D_1 D_2 D_3 D_4 D_5][(dD_6/dt)D_6^{-1}][D_1 D_2 D_3 D_4 D_5]^{-1},\end{aligned} \tag{2.33}$$

or, in terms of velocity matrices,

$$\begin{aligned}[(dD_h/dt)D_h^{-1}] = {} & [V(d\theta_1/dt, ds_1/dt, \mathbf{u}_{10}, \mathbf{Q}_{10})] + \cdots \\ & + [D_1 D_2][V(d\theta_3/dt, ds_3/dt, \mathbf{u}_{30}, \mathbf{Q}_{30})][D_1 D_2]^{-1} + \cdots \\ & + [D_1 D_2 D_3 D_4 D_5][V(d\theta_6/dt, ds_6/dt, \mathbf{u}_{60}, \mathbf{Q}_{60})][D_1 D_2 D_3 D_4 D_5]^{-1}.\end{aligned} \tag{2.34}$$

First note that the velocity matrices in Eq. (2.34) contain the ZRP data $(\mathbf{u}_{io}, \mathbf{Q}_{io})$. Second, note that displacement matrix D_1 changes the second joint axis at ZRP: $(\mathbf{u}_{20}, \mathbf{Q}_{20})$ to the current position: $(\mathbf{u}_2, \mathbf{Q}_2)$, matrix $[D_1 D_2]$ changes $(\mathbf{u}_{30}, \mathbf{Q}_{30})$ to $(\mathbf{u}_3, \mathbf{Q}_3)$, and so on. Therefore, Eq. (2.31) simplifies the above as follows:

$$\begin{aligned}[(dD_h/dt)D_h^{-1}] = {} & [V(d\theta_1/dt, ds_1/dt, \mathbf{u}_1, \mathbf{Q}_1)] \\ & + [V(d\theta_2/dt, ds_2/dt, \mathbf{u}_2, \mathbf{Q}_2)] \\ & + \cdots + [V(d\theta_6/dt, ds_6/dt, \mathbf{u}_6, \mathbf{Q}_6)].\end{aligned} \tag{2.35}$$

Note that the velocity matrices now contain the current joint axis positions $(\mathbf{u}_i, \mathbf{Q}_i)$. Using the matrix representations in Eq. (2.24) for the left-hand side and Eq. (2.29) for the right-hand side, this becomes

$$\begin{bmatrix} \Omega_h & \mathbf{v}_{o'} \\ 0 & 0 \end{bmatrix} = \sum_i \begin{bmatrix} (d\theta_i/dt)U_i & (ds_i/dt)\mathbf{u}_i - (d\theta_i/dt)U_i\mathbf{Q}_i \\ \mathbf{0} & 0 \end{bmatrix}. \tag{2.36}$$

The left-hand matrix contains the velocity data for the hand, i.e., the angular velocity of the hand (matrix Ω_h) and the velocity $(\mathbf{v}_{o'})$ of a point O' in the hand that coincides with the base origin. The right-hand-side sum ($i = 1, 6$) of matrices contains the velocity data for the joints. Depending upon the nature of the joint, either $ds_i/dt = 0$ (revolute joint) or $d\theta_i/dt = 0$ (prismatic joint); however, in absence of the knowledge of the types of joints, both of these terms are carried in the expression. By comparison of terms, we find the following matrix equations, which relate the hand velocity data and the joint velocity data:

$$\begin{aligned} \Omega_h &= \sum_i (d\theta_i/dt) U_i, \\ \mathbf{v}_{O'} &= \sum_i \{(ds_i/dt)\mathbf{u}_i - (d\theta_i/dt) U_i \mathbf{Q}_i\}. \end{aligned} \tag{2.37}$$

These can be written in the vector form as

$$\begin{aligned} \boldsymbol{\omega}_h &= \sum_i (d\theta_i/dt)\mathbf{u}_i, \\ \mathbf{v}_{O'} &= \sum_i \{(ds_i/dt)\mathbf{u}_i + (d\theta_i/dt)\mathbf{Q}_i \times \mathbf{u}_i\}. \end{aligned} \tag{2.38}$$

Finally, assuming each joint as a specific R or P joint, these six equations can be written as

$$\begin{bmatrix} \boldsymbol{\omega}_h \\ \mathbf{v}_{O'} \end{bmatrix} = [J']\{d\mathbf{q}/dt\}, \tag{2.39}$$

where $\{d\mathbf{q}/dt\} = (\ldots, d\theta_i/dt, \ldots, ds_j/dt, \ldots)^t$ is a 6×1 vector of joint rates; the ith joint is assumed to be revolute, and the jth joint is assumed to be prismatic. The 6×6 matrix $[J']$ is called the velocity Jacobian matrix of the manipulator. This is an important equation, which relates the velocity state of the hand to the joint rates. The columns of $[J']$ have typical forms for revolute (ith) and prismatic (jth) joints,

$$[J'] = \begin{bmatrix} \cdots & \mathbf{u}_i & \cdots & \mathbf{0} & \cdots \\ \cdots & \mathbf{Q}_i \times \mathbf{u}_i & \cdots & \mathbf{u}_j & \cdots \end{bmatrix}. \tag{2.40}$$

The 6×1 column corresponding to the revolute joint (ith) contains the current direction of the axis ($\mathbf{u}_i$) and its "moment" about the base origin ($\mathbf{Q}_i \times \mathbf{u}_i$). The first three elements of the column corresponding to the prismatic joint (jth) are zero, and the last three elements contain the current axis direction ($\mathbf{u}_j$). Although simple in appearance, the above form for the Jacobian requires extensive computations because the current values of ($\mathbf{u}_i, \mathbf{Q}_i$) must be used. For example, current values of ($\mathbf{u}_3, \mathbf{Q}_3$) are obtained from the ZRP data ($\mathbf{u}_{30}, \mathbf{Q}_{30}$) as follows:

$$\mathbf{u}_3 = (R_1 R_2)\mathbf{u}_{30},$$
$$\begin{bmatrix} \mathbf{Q}_3 \\ 1 \end{bmatrix} = [D_1 D_2] \begin{bmatrix} \mathbf{Q}_{30} \\ 1 \end{bmatrix}. \tag{2.41}$$

When the Jacobian matrix $[J']$ is nonsingular, there is a one-to-one relation among the velocity state of the hand and the six joint rates. Singularities in manipulator configuration occur when Jacobian $[J']$ becomes singular, i.e., its determinant becomes zero. This also happens when the otherwise distinct position solutions coalesce into fewer solutions, as was discussed in the inverse position solutions. It should be noted that the forward velocity problem (i.e., determining the hand velocity state from given joint rates) is well defined even when the Jacobian matrix is singular. It is the inverse velocity problem (i.e., determining the joint rates from a given velocity state of the hand) that cannot be solved if the Jacobian matrix becomes singular. Another way to state this is that near a singularity, the manipulator can be controlled in the joint space (i.e., the motion of joints without any specifications on the motion of the hand), but not in the three-dimensional task space (in which the hand moves according to specifications). Loosely, we can say that the hand motion cannot be controlled near a singularity.

In Eq. (2.39), the left-hand side contains the velocity data of the hand as its angular velocity ($\boldsymbol{\omega}_h$) and the velocity of a reference point ($\mathbf{v}_{o'}$) in the hand, which is taken as the body point (O') coincident with the base origin (O). As the hand moves, point O' is a different body point of the hand at different times. In order to avoid this constantly changing definition of point O', it is sometimes desirable to use a point P in the hand as a reference point for defining the velocity state of the hand. Recall the relation between the velocities of points P and O', both of which belong to the hand:

$$\mathbf{v}_P = \mathbf{v}_{o'} + \boldsymbol{\omega}_h \times \mathbf{P}. \tag{2.22}$$

Substituting for $\boldsymbol{\omega}_h$ and $\mathbf{v}_{o'}$ from Eqs. (2.38), we get

$$\mathbf{v}_P = \sum_i \{(ds_i/dt)\mathbf{u}_i + (d\theta_i/dt)(\mathbf{Q}_i - \mathbf{P}) \times \mathbf{u}_i\}. \tag{2.42}$$

The complete velocity relation then becomes

$$\begin{bmatrix} \boldsymbol{\omega}_h \\ \mathbf{v}_P \end{bmatrix} = [J]\{d\mathbf{q}/dt\}, \tag{2.43}$$

where the modified form of the velocity Jacobian $[J]$ is defined, with typical ith revolute joint and jth prismatic joint, as

$$[J] = \begin{bmatrix} \cdots & \mathbf{u}_i & \cdots & \mathbf{0} & \cdots \\ \cdots & \mathbf{u}_i \times \mathbf{Q}_i\mathbf{P} & \cdots & \mathbf{u}_j & \cdots \end{bmatrix}, \tag{2.44}$$

where $\mathbf{Q}_i\mathbf{P} = (\mathbf{P} - \mathbf{Q}_i)$. Term $\mathbf{u}_i \times \mathbf{Q}_i\mathbf{P} = \mathbf{P}\mathbf{Q}_i \times \mathbf{u}_i$ can be interpreted as the "moment" of the revolute axis unit vector $\mathbf{u}_i$ with respect to the reference point P of the hand.

Example 2.3. Derive the 6×6 matrix that can be added to (or subtracted from) the Jacobian matrix J' to obtain J.

Solution: When the reference point for finding the Jacobian is changed, only the columns associated with the revolute joints are affected, and even there, only the cross-product terms are affected.

$$\mathbf{u}_i \times \mathbf{Q}_i\mathbf{P} = \mathbf{u}_i \times (\mathbf{P} - \mathbf{Q}_i) = \mathbf{u}_i \times \mathbf{P} + \mathbf{Q}_i \times \mathbf{u}_i.$$

Then, from the definitions of J and J',

$$[J] = [J'] + \begin{bmatrix} \cdots & \mathbf{0} & \cdots & \mathbf{0} & \cdots \\ \cdots & \mathbf{u}_i \times \mathbf{P} & \cdots & \mathbf{0} & \cdots \end{bmatrix}.$$

Typical columns of the adjustment matrix corresponding to the ith revolute joint and jth prismatic joint (all zeros) are shown.

For robots with spherical wrists (i.e., cointersecting wrist axes), simplification of Jacobian $[J]$ is possible if the body point of the hand that is coincident with the spherical wrist center H is used to define the velocity state of the hand (i.e., point P is replaced by point H_7). In this case, points H_4, H_5, H_6, and H_7 are identical. The reason for simplification is that $\mathbf{u}_i \times \mathbf{Q}_i\mathbf{H}_7 = 0$ for $i = 4, 5, 6$, i.e., the "moments" of the wrist axis directions with respect to the cointersection point H vanish. The velocity relations are

$$\begin{bmatrix} \boldsymbol{\omega}_h \\ \mathbf{v}_{H7} \end{bmatrix} = [J'']\{d\mathbf{q}/dt\}, \tag{2.45}$$

where the simplified Jacobian $[J'']$ has the following partitioned form in which each element is a 3×3 matrix,

$$[J''] = \begin{bmatrix} J_1 & J_3 \\ J_2 & 0_{3\times 3} \end{bmatrix}, \tag{2.46}$$

where, assuming a three-revolute jointed regional structure,

$$\begin{aligned} J_1 &= [\mathbf{u}_1 \;\; \mathbf{u}_2 \;\; \mathbf{u}_3] \\ J_2 &= [\mathbf{u}_1 \times \mathbf{Q}_1\mathbf{H} \;\; \mathbf{u}_2 \times \mathbf{Q}_2\mathbf{H} \;\; \mathbf{u}_3 \times \mathbf{Q}_3\mathbf{H}] \\ J_3 &= [\mathbf{u}_4 \;\; \mathbf{u}_5 \;\; \mathbf{u}_6]. \end{aligned}$$

In this form, we were able to see the decoupling that occurs between the velocities of the regional structure joints ($H_4 = H_7$ leads to $\mathbf{v}_{H4} = \mathbf{v}_{H7}$) and wrist structure joints; these joint velocities are identified by subscripts "reg" and "wrist," respectively.

$$\begin{aligned} \mathbf{V}_{H4} &= [J_2]\{d\mathbf{q}/dt\}_{\text{reg}}, \\ \boldsymbol{\omega}_h - [J_1]\{d\mathbf{q}/dt\}_{\text{reg}} &= [J_3]\{d\mathbf{q}/dt\}_{\text{wrist}}. \end{aligned} \tag{2.47}$$

Given the velocity state of the hand as $(\boldsymbol{\omega}_h, \mathbf{v}_P)$, we can find

$$\mathbf{v}_{H4} = \mathbf{v}_P + \boldsymbol{\omega}_h \times \mathbf{PH}.$$

From the first set of equations (2.47), the regional joint rates $\{d\mathbf{q}/dt\}_{\text{reg}}$ can be found, and then from the second set of equations (2.47), the wrist joint rates $\{d\mathbf{q}/dt\}_{\text{wrist}}$ can be found. The determinant of $[J'']$ can be found as a product of two determinants as follows:

$$\det|J''| = -\det|J_2| \cdot \det|J_3|. \tag{2.48}$$

In Eq. (2.48), $\det|J_2|$ is the determinant of the regional structure sub-Jacobian (3×3) and $\det|J_3|$ that of the wrist structure sub-Jacobian (3×3). We can also conclude for this case that the singularities of the regional structure ($\det|J_2| = 0$) are decoupled from those of the wrist structure ($\det|J_3| = 0$).

Example 2.4. Derive the expression for the 3×3 regional structure sub-Jacobian for the Stanford Arm manipulator of Section 2.3 and find its determinant.

Solution: Because the Stanford Arm robot has an equivalent spherical wrist, the Jacobian should be formulated by using the wrist center

H as the reference point to take advantage of decoupling. This 6×6 Jacobian, J'', has the form given in Eq. (2.46). Important 3×3 sub-Jacobians are the regional structure sub-Jacobian J_2 and the wrist structure sub-Jacobian J_3, where

$$J_2 = [\mathbf{u}_1 \times \mathbf{Q}_1\mathbf{H} \ \ \mathbf{u}_2 \times \mathbf{Q}_2\mathbf{H} \ \ \mathbf{u}_3], \quad J_3 = [\mathbf{u}_4 \ \ \mathbf{u}_5 \ \ \mathbf{u}_6].$$

Note that $\mathbf{H} = \mathbf{H}_4 = \mathbf{H}_7$. Furthermore, J_1 in Eq. (2.46) affects the velocity calculations, but it does not have a major significance otherwise. To find J_2, we need the current values of $\mathbf{u}_1$, $\mathbf{u}_2$, $\mathbf{u}_3$, $\mathbf{Q}_1$, $\mathbf{Q}_2$, and $\mathbf{H}$. The joint displacement matrices were defined in Section 2.3. Notice that

$$\mathbf{u}_1 = \mathbf{u}_{1,0} = (0,0,1)^t, \quad \mathbf{Q}_1 = \mathbf{Q}_2 = \mathbf{Q}_{1,0} = \mathbf{Q}_{2,0} = (0,0,0)^t.$$

Therefore, only $\mathbf{u}_2$, $\mathbf{u}_3$, and $\mathbf{H}$ need to be calculated.

$$\begin{bmatrix} \mathbf{u}_2 \\ 0 \end{bmatrix} = D_1 \begin{bmatrix} \mathbf{u}_{2,0} \\ 0 \end{bmatrix} = \begin{bmatrix} -\sin\theta_1 \\ \cos\theta_1 \\ 0 \\ 0 \end{bmatrix}$$

$$\begin{bmatrix} \mathbf{u}_3 \\ 0 \end{bmatrix} = D_1 D_2 \begin{bmatrix} \mathbf{u}_{3,0} \\ 0 \end{bmatrix} = \begin{bmatrix} \cos\theta_1\cos\theta_2 \\ \sin\theta_1\cos\theta_2 \\ -\sin\theta_2 \\ 0 \end{bmatrix}$$

$$\begin{bmatrix} \mathbf{H} \\ 1 \end{bmatrix} = D_1 D_2 D_3 \begin{bmatrix} \mathbf{H}_0 \\ 1 \end{bmatrix} = \begin{bmatrix} -a\sin\theta_1 + s_3\cos\theta_1\cos\theta_2 \\ a\cos\theta_1 + s_3\sin\theta_1\cos\theta_2 \\ -s_3\sin\theta_2 \\ 1 \end{bmatrix}.$$

After evaluating the cross products $\mathbf{u}_1 \times \mathbf{Q}_1\mathbf{H}$ and $\mathbf{u}_2 \times \mathbf{Q}_2\mathbf{H}$ and substituting into the expression for J_2,

$$J_2 = \begin{bmatrix} -a\cos\theta_1 - s_3\sin\theta_1\cos\theta_2 & -s_3\cos\theta_1\sin\theta_2 & \cos\theta_1\cos\theta_2 \\ -a\sin\theta_1 + s_3\cos\theta_1\cos\theta_2 & -s_3\sin\theta_1\sin\theta_2 & \sin\theta_1\cos\theta_2 \\ 0 & -s_3\cos\theta_2 & -\sin\theta_2 \end{bmatrix}.$$

Determinant $\det|J_2| = -(s_3)^2\cos\theta_2$. Also note that $J_2 = [\nabla^t\mathbf{H}]$.

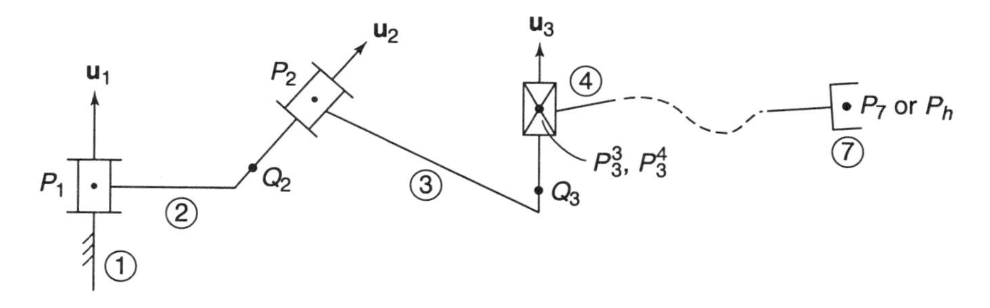

FIGURE 2.9. Manipulator sketch used to derive the manipulator Jacobian from the first principles.

Velocity Analysis from First Principles

Consider, for example, an RRP-3R serial manipulator (Figure 2.9). Let P_i be the point at the center of the ith joint with position vector $\mathbf{P}_i$. Then, starting from the base link (#1), we can analyze each body for velocity relations until the hand body (#7) is reached. The velocity relations are mostly for the case when two points are on the same body, but the coincident point equation is needed when prismatic joints are encountered. The coincident point equation relates the velocities of coincident points on two different bodies. In the example considered, the third joint is prismatic to illustrate this.

For body 2,

$$\begin{aligned} \boldsymbol{\omega}_2 &= (d\theta_1/dt)\mathbf{u}_1, \\ \mathbf{v}_{P_2} &= \boldsymbol{\omega}_2 \times \mathbf{P}_1\mathbf{P}_2 = (d\theta_1/dt)\mathbf{u}_1 \times \mathbf{P}_1\mathbf{P}_2. \end{aligned} \tag{2.49}$$

At prismatic joint 3, the meaning of P_3 is not clear. So, define two coincident points: P_3^3 belonging to body 3, and P_3^4 belonging to body 4; they have the same physical location at an instant but have different velocities. Then, for body 3,

$$\begin{aligned} \boldsymbol{\omega}_3 &= (d\theta_1/dt)\mathbf{u}_1 + (d\theta_2/dt)\mathbf{u}_2, \\ \mathbf{v}_{P_3^3} &= \mathbf{v}_{P_2} + \boldsymbol{\omega}_3 \times \mathbf{P}_2\mathbf{P}_3 \\ &= (d\theta_1/dt)\mathbf{u}_1 \times \mathbf{P}_1\mathbf{P}_3 + (d\theta_2/dt)\mathbf{u}_2 \times \mathbf{P}_2\mathbf{P}_3. \end{aligned} \tag{2.50}$$

Applying the coincident point equation at prismatic joint 3,

$$\mathbf{v}_{P_3^4} = \mathbf{v}_{P_3^3} + \mathbf{v}_{\text{rel}} = \mathbf{v}_{P_3^3} + (ds_3/dt)\mathbf{u}_3$$

or

$$\mathbf{v}_{P_3^4} = (d\theta_1/dt)\mathbf{u}_1 \times \mathbf{P}_1\mathbf{P}_3 + (d\theta_2/dt)\mathbf{u}_2 \times \mathbf{P}_2\mathbf{P}_3 + (ds_3/dt)\mathbf{u}_3. \quad (2.51)$$

For body 4,

$$\begin{aligned} \boldsymbol{\omega}_4 &= (d\theta_1/dt)\mathbf{u}_1 + (d\theta_2/dt)\mathbf{u}_2 + 0\mathbf{u}_3 \\ \mathbf{v}_{P_4} &= \mathbf{v}_{P_3^4} + \boldsymbol{\omega}_4 \times \mathbf{P}_3\mathbf{P}_4 = (d\theta_1/dt)\mathbf{u}_1 \times \mathbf{P}_1\mathbf{P}_4 \qquad (2.52) \\ &\quad + (d\theta_2/dt)\mathbf{u}_2 \times \mathbf{P}_2\mathbf{P}_4 + (ds_3/dt)\mathbf{u}_3. \end{aligned}$$

The general pattern for $\boldsymbol{\omega}_i$ and $\mathbf{v}_{P_i}$ is now established. The velocity state of the hand ($h = 7$) is

$$\begin{aligned} \boldsymbol{\omega}_h &= (d\theta_1/dt)\mathbf{u}_1 + (d\theta_2/dt)\mathbf{u}_2 + 0\mathbf{u}_3 + (d\theta_4/dt)\mathbf{u}_4 \\ &\quad + (d\theta_5/dt)\mathbf{u}_5 + (d\theta_6/dt)\mathbf{u}_6 \\ \mathbf{v}_{P_h} &= (d\theta_1/dt)\mathbf{u}_1 \times \mathbf{P}_1\mathbf{P}_h + (d\theta_2/dt)\mathbf{u}_2 \times \mathbf{P}_2\mathbf{P}_h + (ds_3/dt)\mathbf{u}_3 \\ &\quad + (d\theta_4/dt)\mathbf{u}_4 \times \mathbf{P}_4\mathbf{P}_h + (d\theta_5/dt)\mathbf{u}_5 \times \mathbf{P}_5\mathbf{P}_h \\ &\quad + (d\theta_6/dt)\mathbf{u}_6 \times \mathbf{P}_6\mathbf{P}_h. \qquad (2.53) \end{aligned}$$

Instead of working with the ith joint center P_i (with position vector $\mathbf{P}_i$), we can switch to a general point $\mathbf{Q}_i$ on the ith joint axis. Noting that

$$\mathbf{u}_i \times \mathbf{P}_i\mathbf{P}_h = \mathbf{u}_i \times (\mathbf{P}_i\mathbf{P}_h + c_i\mathbf{u}_i) = \mathbf{u}_i \times \mathbf{Q}_i\mathbf{P}_h$$

and setting $\mathbf{P}_h = \mathbf{P}$, we have

$$\begin{aligned} \boldsymbol{\omega}_h &= (d\theta_1/dt)\mathbf{u}_1 + (d\theta_2/dt)\mathbf{u}_2 + 0\mathbf{u}_3 \\ &\quad + (d\theta_4/dt)\mathbf{u}_4 + (d\theta_5/dt)\mathbf{u}_5 + (d\theta_6/dt)\mathbf{u}_6 \\ \mathbf{v}_P &= (d\theta_1/dt)\mathbf{u}_1 \times \mathbf{Q}_1\mathbf{P} + (d\theta_2/dt)\mathbf{u}_2 \times \mathbf{Q}_2\mathbf{P} + (ds_3/dt)\mathbf{u}_3 \qquad (2.54) \\ &\quad + (d\theta_4/dt)\mathbf{u}_4 \times \mathbf{Q}_4\mathbf{P} + (d\theta_5/dt)\mathbf{u}_5 \times \mathbf{Q}_5\mathbf{P} + (d\theta_6/dt)\mathbf{u}_6 \times \mathbf{Q}_6\mathbf{P}. \end{aligned}$$

These equations correspond to Eqs. (2.43) and (2.44) derived previously from matrix manipulations.

We have so far dealt with the velocity state of the hand (body 7), which is related to the six joint rates through the 6×6 Jacobian matrix, say J. Along similar lines, we can define a $6 \times (k-1)$ link Jacobian J_k for the kth link that relates its velocity state $(\boldsymbol{\omega}_k, \mathbf{v}_{P_k})$ to $(k-1)$ joint rates. The link Jacobian also has columns of the type shown in Eq. (2.44) corresponding to revolute (ith) and prismatic (jth) joints:

$$[J_k] = \begin{bmatrix} \cdots & \mathbf{u}_i & \cdots & \mathbf{0} & \cdots \\ \cdots & \mathbf{u}_i \times \mathbf{Q}_i\mathbf{P}_k & \cdots & \mathbf{u}_j & \cdots \end{bmatrix}, \quad i < k, \; j < k.$$

2.11 Acceleration Relations

The acceleration relations can be found by differentiating the velocity relations. For example, from Eq. (2.43), we get

$$\begin{bmatrix} \boldsymbol{\alpha}_h \\ \mathbf{a}_P \end{bmatrix} = [J]\{d^2\mathbf{q}/dt^2\} + [dJ/dt]\{d\mathbf{q}/dt\}. \tag{2.55}$$

For robots with spherical wrist structure, decoupling also occurs in the acceleration solution, as can be seen from direct differentiation of Eqs. (2.47),

$$\begin{aligned} \mathbf{a}_{H_4} - [dJ_2/dt]\{d\mathbf{q}/dt\}_{\text{reg}} &= [J_2]\{d^2\mathbf{q}/dt^2\}_{\text{reg}}, \\ \boldsymbol{\alpha}_h - [J_1]\{d^2\mathbf{q}/dt^2\}_{\text{reg}} - [dJ_1/dt]\{d\mathbf{q}/dt\}_{\text{reg}} - [dJ_3/dt]\{d\mathbf{q}/dt\}_{\text{wrist}} & \\ = [J_3]\{d^2\mathbf{q}/dt^2\}_{\text{wrist}}. & \end{aligned} \tag{2.56}$$

Thus, in this special case, given the velocity and acceleration states of the hand, we can first solve for the regional joint accelerations and then for the wrist joint accelerations.

2.12 Iterative Position Analysis

Topically, the iterative position analysis should have been discussed right after the analytical inverse position solutions, but it was delayed intentionally because it requires differential analysis quite similar to that done for velocity analysis. We will discuss two methods. The first method is a matrix iteration based upon the Newton–Raphson method. The second method is based upon the solution of nonlinear ordinary differential equations.

Newton–Raphson Method (Matrix Iteration)

Recall the following properties of the joint displacement matrices D_i:

$$D_i = D(\theta_i, s_i, \mathbf{u}_{io}, \mathbf{Q}_{io}) = \begin{bmatrix} R_i & s_i\mathbf{u}_{io} - (R_i - I)\mathbf{Q}_{io} \\ \mathbf{0} & 1 \end{bmatrix}, \tag{2.4}$$

$$(dD_i/dt)D_i^{-1} = \begin{bmatrix} (d\theta_i/dt)U_{io} & (ds_i/dt)\mathbf{u}_{io} - (d\theta_i/dt)U_{io}\mathbf{Q}_{io} \\ \mathbf{0} & 0 \end{bmatrix}$$
$$= [V(d\theta_i/dt, ds_i/dt, \mathbf{u}_{io}, \mathbf{Q}_{io})] \text{ (velocity matrix)} \tag{2.32}$$

or, for the derivative of the joint displacement matrix,

$$(dD_i/dt) = [V(d\theta_i/dt, ds_i/dt, \mathbf{u}_{io}, \mathbf{Q}_{io})]D_i. \tag{2.57}$$

Note the distinct patterns of special constant matrices $B_i = [V(1, 0, \mathbf{u}_{io}, \mathbf{Q}_{io})]$, defined for revolute joints, and $B_j = [V(0, 1, \mathbf{u}_{jo}, \mathbf{Q}_{jo})]$, defined for prismatic joints. Then, $(dD_i/dt) = (d\theta_i/dt)B_iD_i$ for the ith revolute joint, and $(dD_j/dt) = (ds_j/dt)B_jD_j$ for the jth prismatic joint.

The governing equation for joint variables $\mathbf{q} = (\theta_1, \ldots, \theta_i, \ldots, s_j, \ldots, \theta_6)^t$ is

$$D_h = D_1D_2D_3D_4D_5D_6. \tag{2.5}$$

Let the hand trajectory be discretized with respect to time t as $D_h(0)$, $D_h(\Delta T)$, $D_h(2\Delta T), \ldots, D_h(T)$, $D_h(T + \Delta T), \ldots$. Introduce a dummy variable τ: $T \leq \tau \leq T + \Delta T$. Also define

$$\tau + \Delta\tau = T + \Delta T. \tag{2.58}$$

Assume that we have achieved convergence at time T and that the goal is to achieve convergence at the next time $(T + \Delta T)$. The initial guess $(\mathbf{q}_\tau)$ for this iteration will be taken as the joint solution at the previous time T ($\tau = T$, $\Delta\tau = \Delta T$). As the iteration progresses, the joint variable values are $\mathbf{q}_\tau$ at an intermediate dummy time τ, and the dummy time remaining to the goal time is $\Delta\tau = T + \Delta T - \tau$. When the convergence is achieved, $\tau = T + \Delta T$, $\Delta\tau = 0$, and $\Delta\mathbf{q}_\tau = 0$. The dummy time τ should not be confused with real time t when $\tau \neq T$ or $\tau \neq T + \Delta T$; it can be considered as a dummy parameter related to the progression of the Newton–Raphson iteration. From Taylor series expansion:

$$D_h(T + \Delta T) = D_h(\tau + \Delta\tau) \approx D_h(\tau) + [dD_h(\tau)/dt]\Delta\tau + \cdots, \tag{2.59}$$

where $D_h(\tau) = [D_1D_2D_3D_4D_5D_6]_\tau$ and $[dD_h(\tau)/dt]$ are evaluated at dummy time τ with $\mathbf{q}_\tau$. Substituting for D_h and rearranging:

$$D_h(T + \Delta T) - [D_1D_2D_3D_4D_5D_6]_\tau \approx [dD_1/dt]\Delta\tau D_2D_3D_4D_5D_6$$

$$+ D_1[dD_2/dt]\Delta\tau D_3D_4D_5D_6 + \cdots + D_1D_2D_3D_4D_5[dD_6/dt]\Delta\tau. \tag{2.60}$$

Substituting the velocity matrices from Eq. (2.57), introducing constant matrices B_j, and simplifying, we get

$$D_h(T+\Delta T) - [D_1D_2D_3D_4D_5D_6]_\tau \approx [B_1D_1D_2D_3D_4D_5D_6]_\tau \Delta\theta_1$$

$$+ [D_1B_2D_2D_3D_4D_5D_6]_\tau \Delta\theta_2 + \cdots + [D_1D_2D_3D_4D_5B_6D_6]_\tau \Delta\theta_6, \tag{2.61}$$

where $B_i = [V(1, 0, \mathbf{u}_{io}, \mathbf{Q}_{io})]$ for revolute joints and $B_j = [V(0, 1, \mathbf{u}_{jo}, \mathbf{Q}_{jo})]$ for prismatic joints, $(d\theta_i/dt)\Delta\tau = \Delta\theta_i$, $(ds_j/dt)\Delta\tau = \Delta s_j$. These represent twelve equations in the six components of joint variable change vector $\Delta\mathbf{q}_\tau = \{\Delta\theta_1, \Delta\theta_2, \ldots, \Delta\theta_6\}_\tau^t$. Note that $D_h(T + \Delta T)$ does not change during a Newton–Raphson iteration, but all quantities marked with "τ" do change.

One possible approach is to choose six equations by matching the matrix elements: $(1, 4)$, $(2, 4)$, $(3, 4)$, $(3, 3)$, $(2, 3)$, and $(3, 2)$; i.e., the fourth column elements are matched, and from the principal 3×3 minor (rotational part), three elements that are not in the same row or column are matched. Solving the resulting linear system of six equations, $[M]_\tau \Delta\mathbf{q}_\tau \approx \mathbf{b}_\tau$, by using Gaussian Elimination or LU decomposition, we can find the joint variable change $\Delta\mathbf{q}_\tau$. Then, after updating $\mathbf{q}_\tau \leftarrow \mathbf{q}_\tau + \Delta\mathbf{q}_\tau$ $(\tau \leftarrow \tau + \Delta\tau)$, we recompute Eq. (2.61). The meaning of this assignment $(\leftarrow)$ is that we first calculate $\mathbf{q}_{\tau+\Delta\tau} = \mathbf{q}_\tau + \Delta\mathbf{q}_\tau$, and then the calculated $\mathbf{q}_{\tau+\Delta\tau}$ is stored in the memory locations used for $\mathbf{q}_\tau$. This simple device eliminates the need to introduce an iteration counting superscript in formal writing. The dummy times are not calculated or updated, but they are mentioned for clarification only. The convergence is achieved when $|\Delta\mathbf{q}_\tau|$ becomes small enough. When the manipulator is far from any of its singular configurations, this approach is satisfactory. However, near a singularity, the solution becomes very sensitive to the specific choice of six equations. In that case, all twelve equations can be solved as an overdetermined system of linear equations as follows.

Let the twelve linear equations in six unknowns be written as

$$[M]_\tau \Delta\mathbf{q}_\tau \approx \mathbf{b}_\tau, \tag{2.62}$$

where $[M]$ is a 12×6 matrix, $\Delta\mathbf{q}_\tau$ is a 6×1 vector and $\mathbf{b}$ is a 12×1 vector. Premultiply by the transpose of $[M]$ to get a system of six equations in six unknowns as follows:

$$[M^t M]\Delta \mathbf{q}_\tau \approx M^t \mathbf{b}. \tag{2.63}$$

This system can be solved by using Gaussian elimination or LU decomposition, and the solution can be represented symbolically as

$$\Delta \mathbf{q}_\tau \approx [M^t M]^{-1} M^t \mathbf{b}. \tag{2.64}$$

In fact, this is the least-square solution of the overdetermined system of linear Eq. (2.62). Alternatively, as discussed in Section 1.6, the singular value decomposition (SVD), which is slower but more fail-safe, can be used to solve Eqs. (2.62) directly, in the least-square sense. Then, as before, update $\mathbf{q}_\tau \leftarrow \mathbf{q}_\tau + \Delta\mathbf{q}$ ($\tau \leftarrow \tau + \Delta\tau$) and recompute Eq. (2.61). The convergence is achieved when $|\Delta \mathbf{q}_\tau|$ becomes small enough. This approach requires more computations, but the iteration is more stable near singularities.

The iteration can be made more robust and globally convergent by introducing step-size cuts and a strict descent feature for matrix norm $\|D_h(T+\Delta T) - [D_1 D_2 D_3 D_4 D_5 D_6]_\tau\|$, but the related details will not be discussed here.

Improved Matrix Iteration

The matrix iteration of the previous section can be reformulated to utilize the attractive velocity Jacobian $[J']$ in Eq. (2.40). It needs to be computed subsequently for the velocity and acceleration analyses anyway. From the current iteration status (at dummy time τ and joint variable values $\mathbf{q}_\tau$), we start with the first-order estimate of the goal position (at $T + \Delta T$):

$$D_h(T{+}\Delta T) - [D_1 D_2 D_3 D_4 D_5 D_6]_\tau \approx [dD_1/dt]\Delta\tau\, D_2 D_3 D_4 D_5 D_6 \\ + D_1[dD_2/dt]\Delta\tau\, D_3 D_4 D_5 D_6 + \cdots + D_1 D_2 D_3 D_4 D_5 [dD_6/dt]\Delta\tau. \tag{2.60}$$

Except for $D_h(T{+}\Delta T)$, all other quantities are evaluated at dummy time τ with $\mathbf{q}_\tau$. Postmultiplying by $[D_1 D_2 D_3 D_4 D_5 D_6]_\tau^{-1}$, we get an error matrix $[\nu]_\tau$ as follows (Greek symbol ν should not be confused with velocity $\mathbf{v}$ or speed v):

$$[\nu]_\tau = [D_h(T{+}\Delta T)][D_1 D_2 D_3 D_4 D_5 D_6]_\tau^{-1} - [I] \approx [(dD_1/dt)D_1^{-1}]\Delta\tau \\ +[D_1][(dD_2/dt)D_2^{-1}][D_1]^{-1}\Delta\tau + [D_1 D_2][(dD_3/dt)D_3^{-1}][D_1 D_2]^{-1}\Delta\tau$$

$$+\cdots+[D_1D_2D_3D_4D_5][(dD_6/dt)D_6^{-1}][D_1D_2D_3D_4D_5]^{-1}\Delta\tau, \quad (2.65)$$

or, in terms of velocity matrices [Eq. (2.32)],

$$\begin{aligned}[\nu]_\tau \approx\ & [V(d\theta_1/dt, ds_1/dt, \mathbf{u}_{10}, \mathbf{Q}_{10})]\Delta\tau \\ & +\cdots+[D_1D_2][V(d\theta_3/dt, ds_3/dt, \mathbf{u}_{30}, \mathbf{Q}_{30})][D_1D_2]^{-1}\Delta\tau \\ & +\cdots+[D_1D_2D_3D_4D_5][V(d\theta_6/dt, ds_6/dt, \mathbf{u}_{60}, \mathbf{Q}_{60})] \\ & \cdot[D_1D_2D_3D_4D_5]^{-1}\Delta\tau. \end{aligned} \quad (2.66)$$

Making a transition on the right-hand side similar to that from Eq. (2.34) to (2.35), utilizing the velocity similarity principle in Eq. (2.31), the above simplifies as

$$\begin{aligned}[\nu]_\tau \approx\ & [V(d\theta_1/dt, ds_1/dt, \mathbf{u}_1, \mathbf{Q}_1)]\Delta\tau + [V(d\theta_2/dt, ds_2/dt, \mathbf{u}_2, \mathbf{Q}_2)]\Delta\tau \\ & +\cdots+[V(d\theta_6/dt, ds_6/dt, \mathbf{u}_6, \mathbf{Q}_6)]\Delta\tau. \end{aligned} \quad (2.67)$$

Note that the velocity matrices now contain the current joint axis positions. Using the matrix representation in Eq. (2.29), and setting $(d\theta_i/dt)\Delta\tau = \Delta\theta_i$, $(ds_j/dt)\Delta\tau = \Delta s_j$, this becomes

$$\begin{aligned}[\nu]_\tau &= [D_h(T+\Delta T)][D_1D_2D_3D_4D_5D_6]_\tau^{-1} - [I] \\ &\approx \sum_i \begin{bmatrix} \Delta\theta_i U_i & \Delta s_i \mathbf{u}_i - \Delta\theta_i U_i \mathbf{Q}_i \\ \mathbf{0} & 0 \end{bmatrix}_\tau . \end{aligned} \quad (2.68)$$

Depending upon the nature of the joint, either $\Delta s_i = 0$ (revolute joint) or $\Delta\theta_i = 0$ (prismatic joint); however, in the absence of knowledge of the types of joints, both of these terms are carried in the expression. If $\Delta\tau$ is small enough, the 3×3 principal minor of error matrix $[\nu]$ should be nearly skew-symmetric; if not, then the trajectory discretization time step ΔT ($\Delta\tau < \Delta T$) should be reduced. Then, utilizing this near skew-symmetric property, we can pick three rotational equations uniquely and write the following six equations:

$$\begin{aligned}\begin{bmatrix}\nu(3,2)\\ \nu(1,3)\\ \nu(2,1)\end{bmatrix} &\approx \sum_i \Delta\theta_i \mathbf{u}_i \quad (2.69) \\ \begin{bmatrix}\nu(1,4)\\ \nu(2,4)\\ \nu(3,4)\end{bmatrix} &\approx \sum_i \underbrace{\{\Delta s_i \mathbf{u}_i - \Delta\theta_i U_i \mathbf{Q}_i\}}_{\text{(matrix form)}} = \sum_i \underbrace{\{\Delta s_i \mathbf{u}_i + \Delta\theta_i \mathbf{Q}_i \times \mathbf{u}_i\}}_{\text{(column vector form)}} . \end{aligned}$$

Finally, these six equations can be written in terms of joint variable changes $\Delta \mathbf{q}_\tau = (\ldots, \Delta\theta_i, \ldots, \Delta s_j, \ldots)^t$ as follows; the ith joint is revolute and the jth joint is prismatic.

$$\begin{bmatrix} \nu(3,2) \\ \nu(1,3) \\ \nu(2,1) \\ \nu(1,4) \\ \nu(2,4) \\ \nu(3,4) \end{bmatrix}_\tau \approx [J']_\tau \Delta \mathbf{q}_\tau, \tag{2.70}$$

where $[J']$ is the Jacobian matrix defined in Eq. (2.40). This system of equations works well when the manipulator is far from its singularities as well as when it is close to them. Solving this system of six equations by using Gaussian Elimination or LU decomposition, or singular value decomposition (SVD), we can find the joint variable change vector $\Delta \mathbf{q}_\tau$. Then, after updating $\mathbf{q}_\tau \leftarrow \mathbf{q}_\tau + \Delta \mathbf{q}_\tau$ $(\tau \leftarrow \tau + \Delta\tau)$, we recompute Eq. (2.68). The convergence is achieved when $|\Delta \mathbf{q}_\tau|$ becomes small enough. This iteration can also be made more robust and globally convergent by introducing step-size cuts and a strict descent feature for matrix norm $\| [\nu]_\tau \|$, but the related details will not be discussed here.

ODE Solution Method

The idea of solving a system of ordinary differential equations (ODEs) to solve the inverse position problem appears puzzling at first. However, the velocity relations (2.43) can be taken as the appropriate first-order ODEs,

$$\begin{bmatrix} \boldsymbol{\omega}_h(t) \\ \mathbf{v}_P(t) \end{bmatrix} = [J(t)]\{d\mathbf{q}/dt\}, \quad \text{or} \quad \{d\mathbf{q}/dt\} = \mathbf{f}(\mathbf{q}, t). \tag{2.43}$$

The trajectory specifications yield the velocity state [see Eqs. (2.22) and (2.24)] of the hand at time t, $\boldsymbol{\omega}_h(t)$ and $\mathbf{v}_P(t)$, and the initial conditions are $\mathbf{q}(0) = \mathbf{q}_o$. The physical characteristics of the system are contained in the Jacobian matrix $[J(t)]$, and system variables are $\mathbf{q}(t)$. This system of ODEs can be solved for joint variables $\mathbf{q}(t)$ by using Runge–Kutta or predictor–corrector methods. Trajectory discretization with time step ΔT leads to station numbers $1, 2, \ldots, n$, $n+1$, $n+2$, $n+3$, $n+4, \ldots$, etc.

The advantages of this approach are:

(i) The numerical techniques for solving ODEs are highly developed, much more so than those available for solving a system of nonlinear equations. Several books have been devoted to this subject, and many sophisticated computer subroutines are readily available.

(ii) A partial ODE solution is available in the form of trajectory position specification $D_h(t)$. This can be used to monitor the accuracy of the ODE solution and to adjust the time step size. This is a unique feature of the problem at hand because the ODEs are derived from the governing positional equations. In a general ODE problem, ODEs represent the only system description that is available, and assessing the accuracy of the solution is more cumbersome (this is done by determining the sensitivity of the solution to a reduction in the step size).

(iii) Higher computational speed.

(iv) An analog–digital circuit can be designed to implement the solution of Eq. (2.43) in hardware.

(v) Unlike Newton–Raphson iteration, each step of an ODE solution is definitive. That is, we do not have to worry about how many iterations will be needed for convergence. If the time step has been selected appropriately, then each step of the Runge–Kutta or predictor–corrector methods produces a solution after a finite amount (and duration) of computations. This becomes desirable if a customized VLSI chip is designed to function as a real-time, general-purpose inverse kinematics processor.

(vi) The joint velocities are available as a by-product of the ODE solution, and if the LU decomposition is used (and saved) to solve the system in Eq. (2.43), it can be reused in acceleration analysis.

The main disadvantage is that an accurate initial solution is required. Also, if predictor–corrector methods are used, which are not self-starting, then several accurate solutions may be required in the beginning to start the procedure. Floating-point computations should be carried out in double precision (or with twelve or more significant digits) to avoid significant buildups of rounding and truncation errors in trajectories of long duration.

Other helpful hints are to calculate the value of π internally as $\pi = 4$ arctan (1.0) and to renormalize all of the input data for axis directions ($\mathbf{u}_{io}$). Predictor–corrector methods are generally faster than Runge–Kutta methods. In predictor–corrector methods, good results are obtained by following the prediction with only one correction [see also item (v) above]. If the monitored positional error [see item (ii) above] for the hand is not acceptable, then it is better to cut the step size rather than to execute many corrector loops. This is because most of the benefit from the corrector occurs in its first application; subsequent applications produce marginal or little improvement and ultimately trigger the step cut if the monitored positional error is not acceptable after the first correction. If the monitored positional error is acceptable after the prediction, then the corrector step can be skipped to further speed up the algorithm. The use of fourth-order formulas for Runge–Kutta and predictor–corrector (Adams–Moulton) is quite satisfactory for robotics applications. Higher-order formulas can be used, but their benefit is questionable from a theoretical point of view. This is because, in many practical cases, the specified hand trajectory itself has continuity only up to the second derivatives (i.e., acceleration). Theoretically, the fourth-order formulas require continuity up to the fifth derivative, and they can be considered sophisticated enough for most practical robotics applications.

The fourth-order Adams–Moulton predictor–corrector formulas are used as follows (note $\dot{\mathbf{q}} = d\mathbf{q}/dt$):

Predictor:

$$\mathbf{q}_{n+4} = \mathbf{q}_{n+3} + \left(\frac{\Delta T}{24}\right)(55\dot{\mathbf{q}}_{n+3} - 59\dot{\mathbf{q}}_{n+2} + 37\dot{\mathbf{q}}_{n+1} - 9\dot{\mathbf{q}}_{n}). \quad (2.71)$$

The predictor formula extrapolates for joint values $\mathbf{q}$ at station $(n+4)$ by using the known solutions for the four previous stations n, $n+1$, $n+2$, and $n+3$. Evaluation (symbolically) of $d\mathbf{q}/dt$ at station $n+4$ using the predicted result is

$$\dot{\mathbf{q}}_{n+4} = [J^{-1}]\begin{bmatrix}\boldsymbol{\omega}_h \\ \mathbf{v}_P\end{bmatrix}. \quad (2.72)$$

Corrector:

$$\mathbf{q}_{n+4} = \mathbf{q}_{n+3} + \left(\frac{\Delta T}{24}\right)(9\dot{\mathbf{q}}_{n+4} + 19\dot{\mathbf{q}}_{n+3} - 5\dot{\mathbf{q}}_{n+2} + \dot{\mathbf{q}}_{n+1}). \quad (2.73)$$

The corrector formula refines the joint values $\mathbf{q}$ at station $n+4$ by using the known solutions for three prior stations $n+1$, $n+2$, and $n+3$ and the evaluated estimation of $d\mathbf{q}/dt$ at the current station $n+4$. Evaluation (symbolically) using the corrected result is

$$\dot{\mathbf{q}}_{n+4} = [J^{-1}] \begin{bmatrix} \boldsymbol{\omega}_h \\ \mathbf{v}_P \end{bmatrix}. \tag{2.74}$$

This is the predictor formula followed by one application of the corrector formula. The time-consuming steps are the evaluation steps. If more corrections are desired, although not recommended here, then Eqs. (2.73) and (2.74) are used repeatedly. It can be seen from these formulas that solution for four previous steps is needed to find the solution for the next step. The initial four points must be found accurately from a lower-order formula (e.g., Euler's formula) or Runge–Kutta method or the iterative Newton–Raphson method. Finally, it should be noted that even though these formulas use only the first derivatives, the fact that these are fourth-order formulas means that piecewise continuous approximations, with fourth-order segments, are produced for the joint trajectories $\mathbf{q}(t)$.

Hybrid Schemes

Hybrid schemes can combine the speed and efficiency of the ODE solution method with the precision of the improved matrix iteration. The predictor of Eq. (2.71) is coupled with the matrix iteration of Eq. (2.70) instead of the evaluation–corrector steps of (2.72)–(2.74). The predictor provides an excellent initial guess for the matrix iteration, which then provides a fast and accurate quadratic convergence to the solution. Other predictors can be substituted for Eq. (2.71).

It should be noted that hybrid schemes are possibly only when the ODEs (e.g., the velocity relations in robotics) are derived from explicit positional equations. The analytical formulations of many physical problems lead directly to ODEs only, and this hybrid device is not possible in those situations; the accuracy of the ODE solution must then be assessed by comparing the intermediate solutions with two different step sizes. The situations where these hybrid schemes can be used do not arise often but include an important class of problems that require the solution of a system of nonlinear equations with one free parameter. Interested readers should consult the literature on continuation methods for further information.

2.13 Problems

1. For a robot manipulator with cylindrical-three-roll configuration (see Figure P1.1), describe the ZRP data assuming that $\mathbf{H}_o = (0, 0, 0)^t$ and $\mathbf{P}_o = (h, 0, 0)^t$. Derive the formulas for its inverse kinematic (position) analysis.

2. For manipulator case 2 in Section 2.4, derive the expression for joint angle θ_2.

3. For manipulator case 3 in Section 2.5, derive the expression for joint angle θ_1.

4. For the PUMA manipulator in Section 2.6, derive the expression for joint angles θ_1 and θ_2.

5. Draw kinematic sketches of the PUMA robot at its three singularity positions.

6. For a PUMA robot, $a = 5''$, $b = c = 20''$, $h = 10''$ (ZRP data table, Section 2.6). The current position of the hand is $\mathbf{P} = (20'', 15'', 0)^t$, $\mathbf{u}_a = (0, 1, 0)^t$, and $\mathbf{u}_t = (0, 0, 1)^t$. There are a total of eight distinct solutions for the joint angles. Of these, find the four solutions for which the joint angle $\theta_1 = 0$.

7. For the wrist solution for the PUMA manipulator in Section 2.7, verify that

   (a) $\mathbf{v}'_a = \mathbf{v}_a$,

   (b) $w'_{tx} = w_{tx}$, $w'_{ty} = -w_{ty}$, $w'_{tz} = -w_{tz}$.

8. For the Cincinnati Milacron T^3 robot in Section 2.8, derive the expressions for joint angles θ_2 and θ_6.

9. Draw kinematic sketches of the Cincinnati Milacron T^3 robot at its three singularity positions.

10. (a) For the robot configuration shown in its ZRP position in Figure P2.1, write the joint matrices $D_i = D(\theta_i, s_i, \mathbf{u}_{io}, \mathbf{Q}_{io})$, $i = 1, 2, 3$.

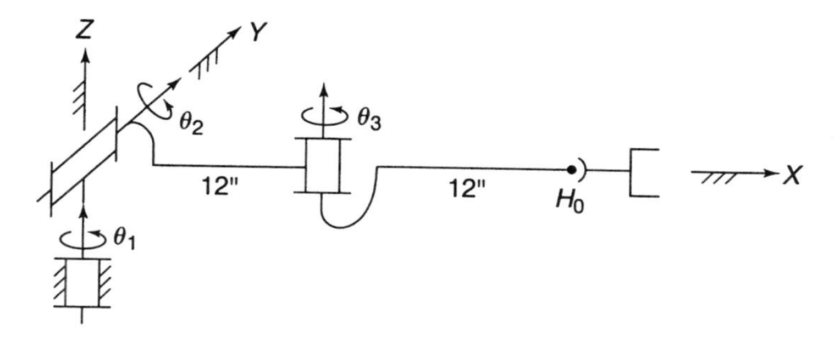

FIGURE P2.1. ZRP position of the manipulator with spherical wrist equivalent for Problem 10.

(b) Derive the following expressions for the coordinates of the wrist center H:

$$\begin{aligned} H_x &= 12(1+\cos\theta_3)\cos\theta_2\cos\theta_1 - 12\sin\theta_3\sin\theta_1 \\ H_y &= 12(1+\cos\theta_3)\cos\theta_2\sin\theta_1 + 12\sin\theta_3\cos\theta_1 \\ H_z &= -12(1+\cos\theta_3)\sin\theta_2. \end{aligned}$$

Derive the stepwise formulas for finding angles θ_1, θ_2, θ_3. (*Hint*: Start with $\mathbf{H}\cdot\mathbf{H}$).

11. The ZRP data for an RPR regional structure is as shown in Figure P2.2. Note that when point H_o is at $(2l, 0, 0)$, then the joint variable s_2 is defined to be zero.

 (a) For general values of θ_1, s_2, and θ_3, find the expressions for the coordinates of point H: (H_x, H_y, H_z).

 (b) Find the expressions for θ_1, s_2, and θ_3 so that these can be calculated in a sequential stepwise fashion from the values of (H_x, H_y, H_z).

 (c) How many distinct solution sets exist for θ_1, s_2, and θ_3? Can all of these be physically realized?

 (d) The singularities of the regional structure are when the point H lies on the Z axis and when $\theta_3 = \pm 90°$. Explain in mathematical *and* physical terms the difficulties at these configurations.

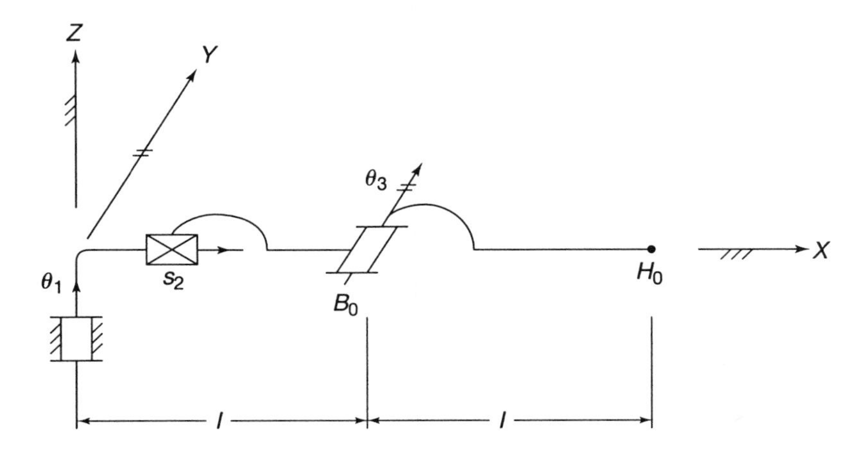

FIGURE P2.2. ZRP position for the manipulator in Problem 11.

12. A robot has spherical (RRP) regional structure and spherical (3R) wrist. The ZRP data are as shown in Figure P2.3, with H_o taken at $(0, 0, 0)$—read the entire problem statement below before starting with the solution.

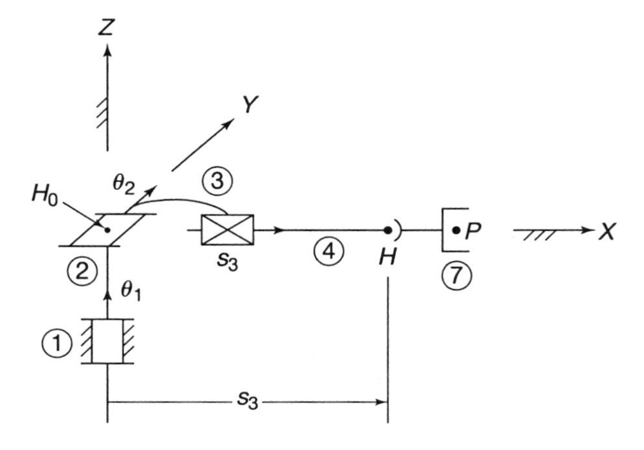

FIGURE P2.3. ZRP position for the manipulator in Problem 12, except that $s_3 \neq 0$ is shown for clarity. Point H is shown in its current position as well as its zero reference position at the origin (H_o). Additional wrist details are not needed for this problem.

(a) Write the general form (only) of the 6×6 Jacobian matrix J'', which is with respect to the wrist center H. Identify the form of the 3×3 regional structure sub-Jacobian J_2.

(b) From the general expressions for the needed $\mathbf{u}_i$, $\mathbf{H}$, etc., derive the general expression for the elements of the 3×3 sub-Jacobian J_2. Compare the result with $[\nabla^t \mathbf{H}]$.

(c) Find the determinant of J_2 and interpret the physical meaning of the conditions for the vanishing of this determinant.

13. Consider a 2-dof serial robot with ZRP data: $\mathbf{u}_{10}$, $\mathbf{Q}_{10}$, $\mathbf{u}_{20}$, $\mathbf{Q}_{20}$, and the following two sequences of joint motions.

    (a) The first rotation is by θ_1 about the first joint axis at ($\mathbf{u}_{10}$, $\mathbf{Q}_{10}$). This rotation, $D(\theta_1, 0, \mathbf{u}_{10}, \mathbf{Q}_{10})$, also displaces the subsequent (i.e., the second) axis from ($\mathbf{u}_{20}$, $\mathbf{Q}_{20}$) to ($\mathbf{u}_2$, $\mathbf{Q}_2$). The second rotation is by θ_2 about this "new" second axis at ($\mathbf{u}_2$, $\mathbf{Q}_2$). Write the resulting displacement of the hand, D_h', by using the active representation.

    (b) The first rotation is by θ_2 about the second axis at ($\mathbf{u}_{20}$, $\mathbf{Q}_{20}$). This rotation, $D(\theta_2, 0, \mathbf{u}_{20}, \mathbf{Q}_{20})$, does not disturb the preceding (i.e., the first) axis from ($\mathbf{u}_{10}$, $\mathbf{Q}_{10}$). The second rotation is by θ_1 about the unperturbed first axis at ($\mathbf{u}_{10}$, $\mathbf{Q}_{10}$). Write the resulting displacement of the hand, D_h'', by using the active representation.

    (c) By using the displacement similarity principle mentioned in Problem 17 of Chapter 1, show that $D_h' = D_h''$. While it appears intuitively that this result may be valid, it is not easy to prove mathematically without the application of the aforementioned displacement similarity principle. This result can be generalized recursively to prove the validity of the ZRP method described in Section 2.2.

14. For the orthogonal bevel-geared wrist, derive the inverse relations in Eq. (2.16).

15. In an oblique bevel-geared wrist, gears G_1, G_3, G_4, and G_6 are identical. The cone angle of gear G_5 is 180°. The oblique angle $\beta = 20°$. Find the half pitch cone angles for all gears in the wrist.

16. To simplify production and service, an oblique bevel-geared three-roll wrist has identical bevel gears G_1, G_3, G_4, and G_6.

The large gear G_5 has a cone angle of 180°. If the oblique angle β is 30°, determine all of the cone angles, and find the matrix that relates the three wrist variables to the three actuation variables, and vice versa.

17. The velocity state of a robot hand is as follows:

$$\boldsymbol{\omega} = 2\mathbf{I} + 5\mathbf{J} + \mathbf{K}\,(\text{rad/sec}) \text{ and } \mathbf{v_P} = 10\mathbf{I} - 3\mathbf{J} + 5\mathbf{K}\ (\text{in/sec}).$$

The coordinates of the reference point P on the hand are (1″, 3″, 4″). Determine the 4×4 velocity matrix $[V]$ with these data.

18. The velocity state of a robot hand is given as

$$\omega = 10 \text{ rad/sec},\ v_{\text{ISA}} = 2 \text{ in/sec},$$

$$\mathbf{u}_{\text{ISA}} = (4/57, 23/57, 52/57)^t,\ \mathbf{Q}_{\text{ISA}} = (2'', 3'', 0)^t.$$

(a) Find the angular velocity matrix $[\Omega]$ and the velocity matrix $[V]$ for the hand.

(b) Find the velocities of points P: $(1'', 1'', 4'')$, (2.4″, 5.3″, 5.2″), and $(0, 0, 0)$. Discuss the results for the second point, which is on ISA, and the third point, which is the body point coincident with the base origin.

19. (a) Derive Eq. (2.32) for the joint velocity matrix by direct substitution from Eq. (2.4). How can this result also be obtained by inspection from Eq. (2.29)?

(b) What are the forms and physical interpretations of the constant matrices $B_i = [V(1, 0, \mathbf{u}_{io}, \mathbf{Q}_{io})]$ and $B_j = [V(0, 1, \mathbf{u}_{jo}, \mathbf{Q}_{jo})]$?

20. For manipulator case 2 in Section 2.4, derive the expression for the determinant of the 3×3 regional structure sub-Jacobian J_2 in Eq. (2.46).

21. For manipulator case 3 in Section 2.5, derive the expression for the determinant of the 3×3 regional structure sub-Jacobian J_2 in Eq. (2.46).

22. A flow chart is a preliminary step in writing complex computer programs. Prepare a flow chart for the Improved Matrix Iteration for inverse position analysis (Section 2.12) featuring Eqs. (2.68) and (2.70).

23. Given $x(t) = 1 + t + t^2 + t^3 + t^4$. Calculate x and (dx/dt) by direct substitution at $t = 1.0$, 1.1, 1.2, and 1.3. Now use the Adams predictor formula [Eq. (2.71)] to estimate $x(1.4)$ and compare this value with that obtained from direct substitution of $t = 1.4$ in the given function $x(t)$. Use six or more decimal places to carry out these calculations.

3
Robot Workspace

3.1 Background

Gross motion capabilities of robot manipulators can be described with respect to the reaches of points, lines, or planes attached to the robot hand. In this section, we will focus upon the point workspaces. On a simple level, the workspaces can be used to establish safe zones around the robot to prevent injury to humans and damage to other machinery on the factory floor. However, qualitative and quantitative studies of workspaces can (i) yield useful insights about the kinematic structure of the robot in the design stage, (ii) lead to criteria for the evaluation of different types of robot arms, and (iii) assist in the planning of desired tasks in favorable zones.

3.2 Workspace Classification

The total workspace of an industrial robot is defined as the space reachable by a reference point in the hand. The normalized workspace should be as large as possible, and the points within the workspace should be reachable in multiple ways to ensure maximum flexibility in planning tasks.

The primary workspace is defined as the subset of the total workspace that contains points around which all orientations of the hand are possible. If the hand has an identifiable axial direction ($\mathbf{u}_a$), then with the reference point of the hand held fixed at a point in the primary workspace, say, by means of an imaginary ball-and-socket connection, it should be possible to orient the axial direction $\mathbf{u}_a$ arbitrarily and to spin the hand freely about the axial direction. In the discussion of workspaces, we consider the idealized situation in which there are no mechanical interferences due to link collisions, joint stops, or physical features of joints such as ball-and-socket joints. The capabilities of a robot in the primary workspace are very useful for intricate tasks and assembly operations. Therefore, it is desirable that a large fraction of the total workspace be the primary workspace. The primary workspace is also referred to as the dexterous workspace in the literature.

The secondary workspace is the remainder of the total workspace. The orientational capabilities of the hand are limited in the secondary workspace. Either the axial direction of the hand cannot be placed in an arbitrary direction with respect to points in the secondary workspace, or the hand cannot spin freely about the achievable axial directions.

The collection of achievable axial directions, with or without full spin around them, at a point in the workspace is called the approach directions. Along an approach direction, the distance by which the axial vector of the hand can slide away from the point in question is called the approach length. By marking approach lengths on all permissible approach directions, an approach envelope (or ray graph) can be found for each point in the workspace (Figure 3.1). Such approach envelopes surround the points in the primary workspace completely [Figure 3.1(a)], but in the secondary workspace, they have singularities at the reachable point itself [Figure 3.1(b)]. A database of this type of workspace feature can be created and stored for future use in task planning.

A general serial robot is shown in Figure 3.2(a). If an n-axis robot has a wrist with three cointersecting revolute joints (R_{n-2}, R_{n-1}, R_n) as shown in Figure 3.2(b), then it can be modeled kinematically by a spherical joint (or ball-and-socket joint) at the cointersection point (or wrist center) H; the wrist bodies $(n-1)$ and n are "eliminated", and the hand body $(n+1)$ is imagined to be "attached" to body $(n-2)$ directly through a ball-and-socket joint at the wrist center H

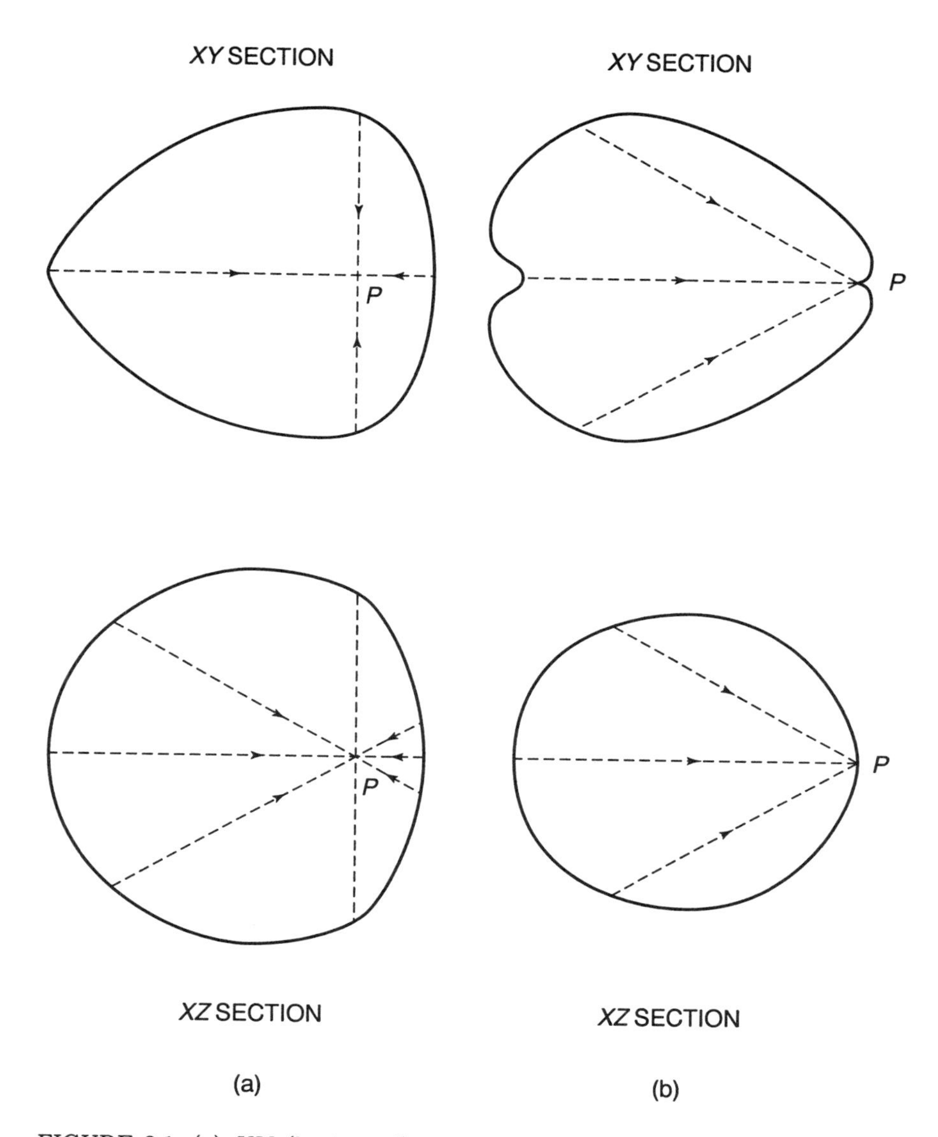

FIGURE 3.1. (a) XY (horizontal) and XZ (vertical) sections of the approach length ray graph for a point in the primary workspace. (b) XY (horizontal) and XZ (vertical) sections of the approach length ray graph for a point in the secondary workspace.

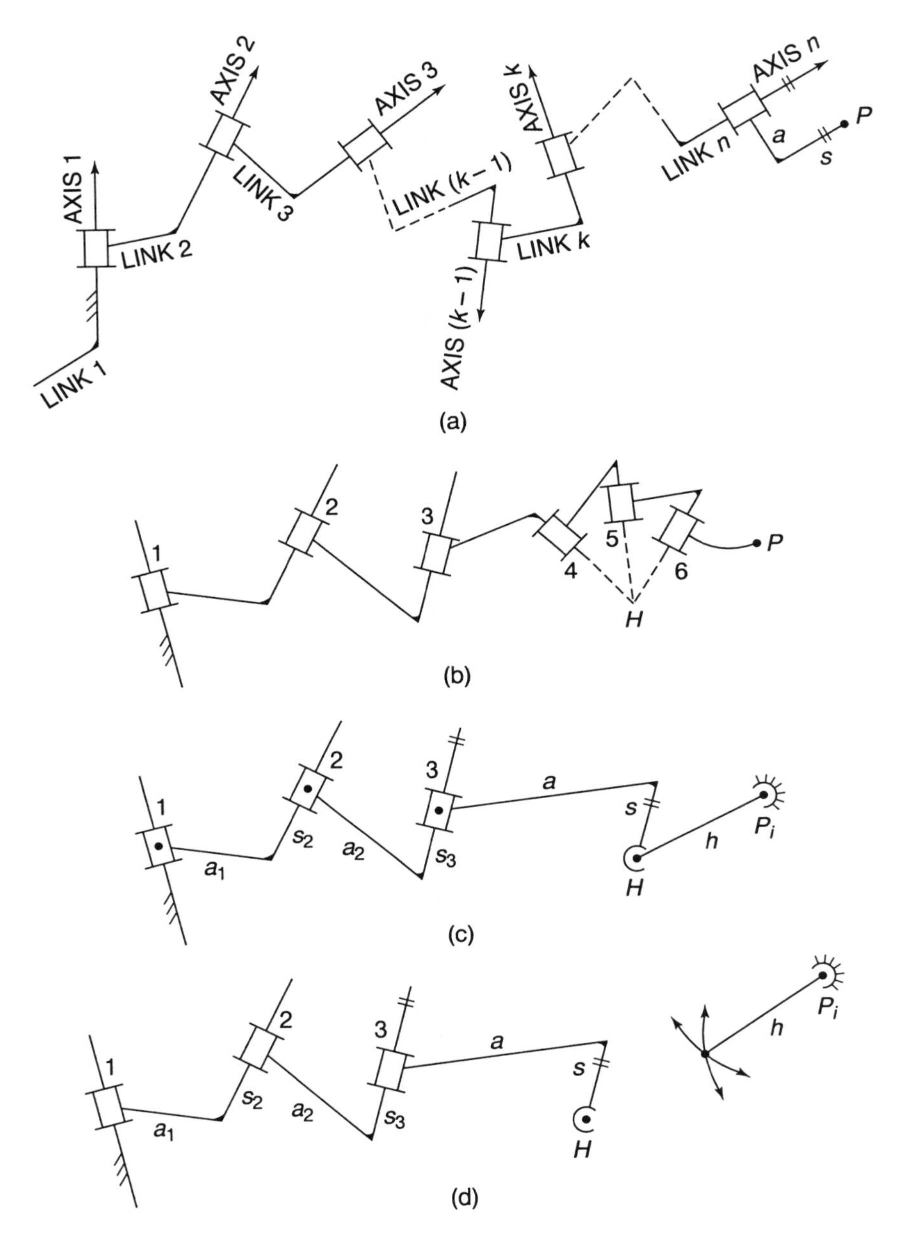

FIGURE 3.2. (a) A general serial N-revolute jointed manipulator. (b) A 6-R manipulator with three cointersecting wrist axes (spherical wrist equivalent). (c) Replacement of the spherical wrist equivalent by the ball-and-socket joint at the wrist center H. Point P_i has been grounded for analyzing the properties of the workspace $W_1(P)$. (d) If point P_i does not belong to the workspace $W_1(P)$, then the ball-and-socket joint cannot be assembled at H.

[Figure 3.2(c)]. If the wrist angles are 90° (i.e., $\alpha_{n-2} = \alpha_{n-1} = 90°$), then the orientational coverage of the nth revolute axis (or the last wrist axis) with respect to the $(n-2)$th revolute axis (or the first wrist axis) is ideally unrestricted. Figure 3.2(d) shows a break at the wrist center H; as discussed later in Section 3.4, this will have a significant implication on par with the decoupling of the regional and wrist structures that we saw in position, velocity, and acceleration solutions. If the hand axial direction $\mathbf{u}_a$ is collinear with the nth revolute joint, then the free spin of the hand about any axial direction is guaranteed. If the hand axis is not collinear with the nth revolute axis but passes through the wrist center H, then the full spin of the hand about the axial direction is not assured. This problem will be considered in Section 3.6.

3.3 Nature of Workspace

Let us discuss the process of generation for the workspace of an n-revolute jointed robot, starting from the last joint (n). Assuming the general case when the reference point P of the hand is offset from the last joint axis (n), we can define the workspace of point P when the nth joint is moved through 360° while all the other joints are kept locked. We will use the notation $W_n(P)$ to represent this workspace of P with respect to axis n, which is a circle (curve) shown in Figure 3.3(a).

Next, the $(n-1)$th joint is unlocked, and it revolves the previous workspace $W_n(P)$ by 360° around the axis $(n-1)$ to produce the next workspace $W_{n-1}(P)$, which is a torus (surface) shown in Figure 3.3(b). Depending upon the relative placement of the circle and the axis $(n-1)$, the torus can be a right-circular torus (i.e., surface of a donut), a sphere, a flattened disk, or a general torus. A general torus can have a hole around the axis of revolution [Figure 3.3(c)]. A hole is an empty space such that we can see through it, or we can pass a line through the hole that does not intersect with the workspace. There can be two types of internal empty spaces (or voids): toroidal and central. The voids are not visible from the outside but can be seen in cross-sectional views. We cannot pass a line through the voids without intersecting the workspace. The toroidal void does not contain the axis of revolution, e.g., the hollow of a right-circular torus. The central void contains the axis of revolution, e.g., when

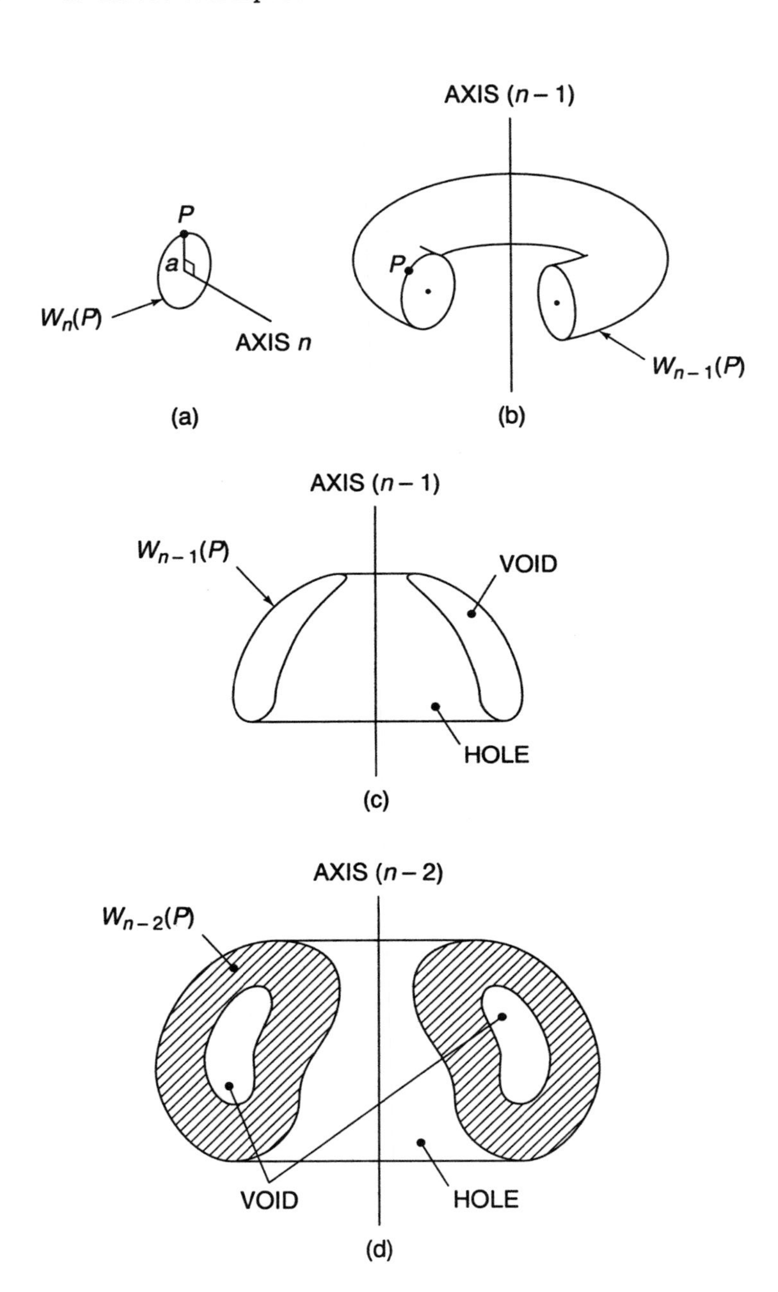

FIGURE 3.3. (a) Generated circle about axis n. (b) Generated right-circular torus about axis $(n-1)$. (c) General torus generated about axis $(n-1)$ with hole and void. (d) Generated complex solid about axis $(n-2)$ with hole and void.

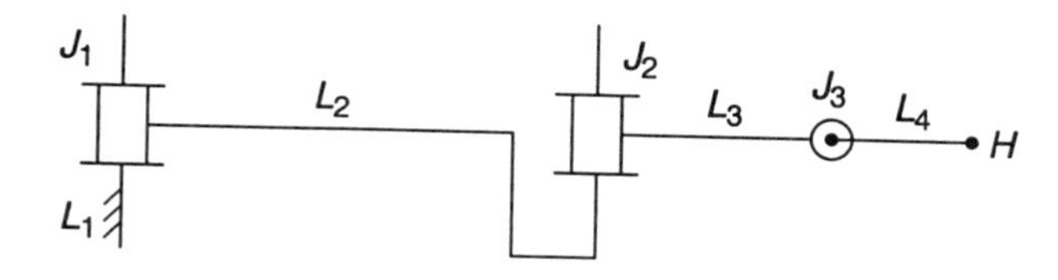

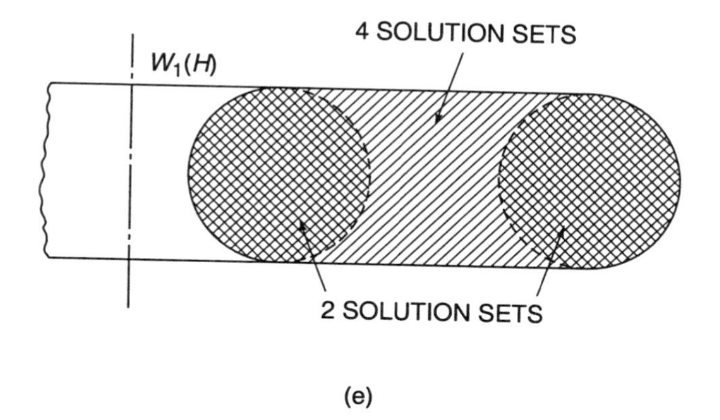

(e)

FIGURE 3.3. (e) Interior and exterior singularity surfaces in the workspace $W_1(H)$.

the circle is revolved around an axis that lies in the plane of the circle and also cuts the circle.

Then, the $(n-2)$th joint is unlocked, and the workspace $W_{n-1}(P)$ is revolved around the axis $(n-2)$ to obtain the next workspace $W_{n-2}(P)$, which is a complex solid of revolution [Figure 3.3(d)]. This complex solid can have a hole, a toroidal void, or central voids; in axial cross-section, containing the axis of revolution, these have irregular boundaries. From this point onwards, the workspaces $W_{n-3}(P)$, $W_{n-4}(P), \ldots, W_1(P)$ have the same topological structure.

Figure 3.3(e) shows the cross-section of workspace $W_1(H)$ of a 3-axis regional structure with revolute axis configuration $R1 \parallel R2 \perp R3$. The cross-hatched zone shows the space in which the regional structure has two distinct position solutions. In the single-hatched zone, four distinct position solutions exist. The surfaces across which the number of distinct solutions change from 4 to 2, or 2 to 0, are singularity surfaces. This example shows that singularity surfaces can be the boundary surfaces or the interior surfaces of the workspace.

Workspace $W_1(P)$ of an n-revolute jointed robot arm is a complex solid that can have holes, toroidal voids, and central voids. Sometimes, the holes or central voids can be designed so that they contain the base column; Stanford Arm and PUMA robots, with offsets at shoulders, are examples of the former, while some ASEA robots are examples of the latter. A toroidal void is always within the active workspace, and it is undesirable in general purpose manipulators.

If the workspace $W_{k+1}(P)$ has a hole, then this hole may change into another hole, toroidal void, central void, or solid in the workspace $W_k(P)$. If $W_{k+1}(P)$ has a toroidal void, then this can change into another toroidal void, central void, or solid in $W_k(P)$. If $W_{k+1}(P)$ has a central void, then it can change into another central void or a solid in $W_k(P)$. Some of the corresponding conditions are identified in Table 3.1.

If the revolute joints have limited rotational movements, then the workspaces produced at each step of generation are incomplete solids of revolution. If some of the joints are prismatic, then the generation procedure is modified. If the kth joint is a prismatic joint with a finite range, then the workspace $W_{k+1}(P)$ is translated with respect to the kth joint to produce a prismatic body for the workspace $W_k(P)$.

3.4 Determination of Primary Workspace

The determination of primary workspace for a general manipulator is a difficult problem. However, for manipulators with a spherical wrist equivalent, this can be done as follows.

Consider a 6R manipulator with a spherical wrist equivalent [Figure 3.2(b)]. It can be modeled as a 3R-S manipulator with spherical joint S placed at wrist center H. Let P be the reference point of the hand. In order to determine whether a specific position P_i of point P belongs to the primary or secondary workspace, we attach point P to the ground through another spherical joint at P_i. Then, the possible directions at which the manipulator can reach point P_i are the same as the allowable directions of the link HP_i in the 3R-2S closed-loop linkage [Figure 3.2(c)]; note that free spin of the hand body about line HP_i is possible. If P_i is outside of the workspace $W_1(P)$, then the 3R-2S linkage cannot be assembled [Figure 3.2(d)].

It can be seen that all orientations of HP_i are possible when the sphere of radius h with center at P_i lies completely within the workspace $W_1(H)$ of the wrist center H [Figure 3.4(a)]. Free spin

TABLE 3.1. Transformations of holes and voids.

In $W_{k+1}(P)$	Dist $(k, k+1)$	Angle $(k, k+1)$	Additional condition relative to $W_{k+1}(P)$	In $W_k(P)$
hole			k through hole, no intersections	hole
hole		small	k does not pass through the hole	solid
hole		large	k intersects, passes almost centrally through the hole	central void
hole		large	k does not pass through the hole	toroidal void
hole		large	k intersects, passes off-centered through the hole	central and toroidal voids
central void			k passes through the central void	central void
central void			k does not pass through the central void	solid
toroidal void			k passes through the toroidal void	central void
toroidal void	small	small		toroidal void
toroidal void			k is outside	solid
toroidal void		large	k is not outside	solid and central void

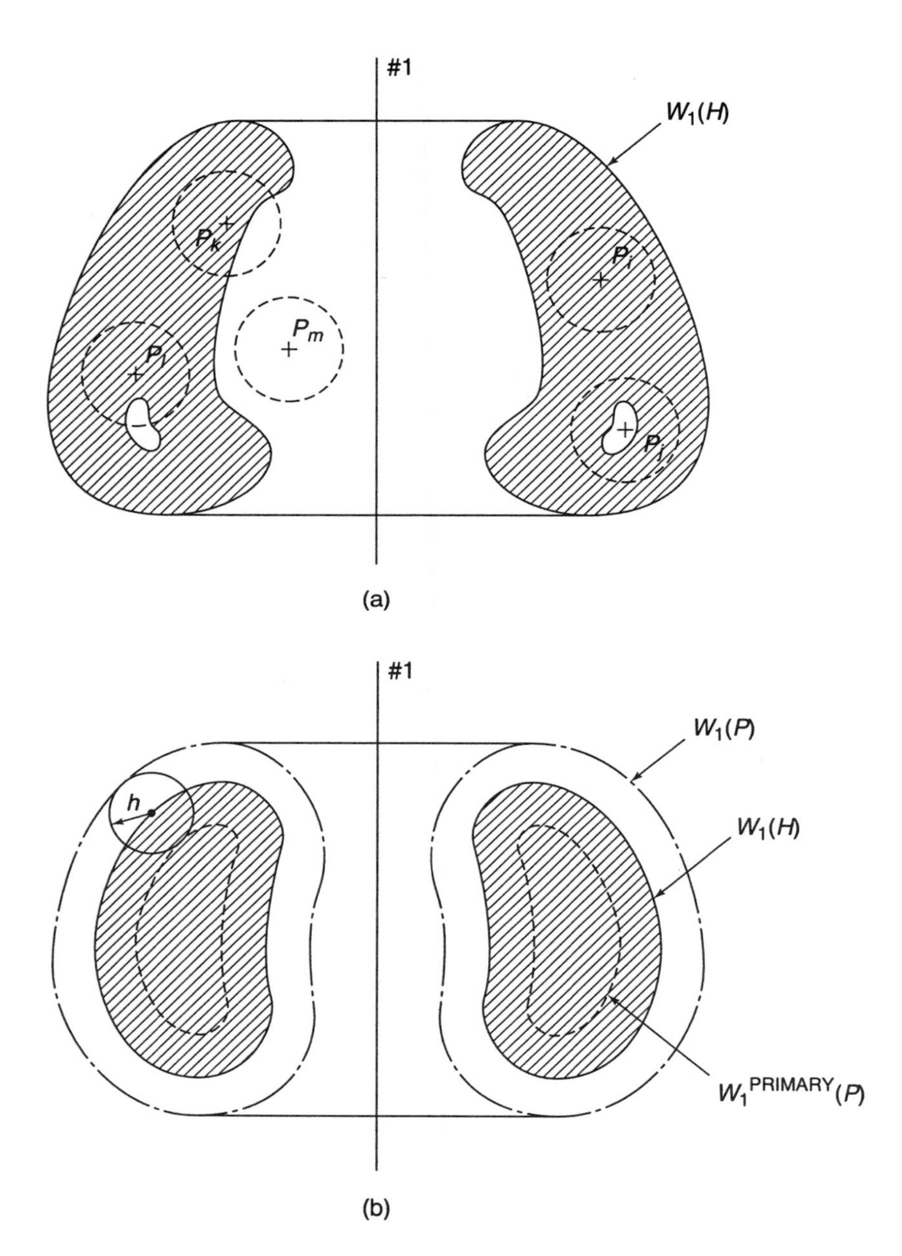

FIGURE 3.4. (a) Properties of workspace $W_1(P)$ derived from workspace $W_1(H)$. (b) Total and primary workspaces for point P obtained from workspace $W_1(H)$ and hand of size h.

is also possible about HP_i. Therefore, all orientations of the hand around P_i are then possible, and point P_i lies within the primary workspace $W_1^p(P)$. It is interesting to note the critical role that workspace $W_1(H)$ plays in this determination. In Figure 3.4(a), it can be seen that point P_m does not lie in the workspace $W_1(P)$; points P_j, P_k, and P_ℓ lie in the secondary workspace $W_1^s(P)$ but not in the primary workspace $W_1^p(P)$; and point P_i lies in the primary workspace $W_1^p(P)$. If workspace $W_1(H)$ does not have internal voids, then a simple way to determine primary and total workspaces is as follows [Figure 3.4(b)]. A sphere of radius h is moved such that its center lies on the boundary of $W_1(H)$; then the inner envelope of these spheres is the boundary of the primary workspace $W_1^p(P)$, and the outer envelope is the boundary of the total workspace $W_1(P)$. If $W_1(H)$ has internal voids, and h is relatively small compared to the void, then the same process can be repeated at the inner boundary of $W_1(H)$, but now the outer envelope is the boundary of $W_1^p(P)$ and the inner envelope is the boundary of $W_1(P)$.

It can also be seen that when hand size $h = 0$, the primary and total workspaces are identical, i.e., all of the total workspace is primary workspace. However, h cannot be zero due to the limitations of the wrist hardware. A nonzero h reduces the primary workspace but increases the total workspace. Therefore, hand size h and overall wrist dimensions should be as small as possible.

For finding the volume of a solid of revolution, multiply the generating cross-sectional area with the distance traveled by the centroid of the area (Theorem of Pappus).

Example 3.1. A robot arm with $R1 \perp R2 \parallel R3$ regional structure and a spherical-equivalent 3R wrist is shown in Figure 3.5(a). Analyze the nature of total and primary workspaces for different values of hand size h.

Solution: The workspace $W_3(H)$ is a circle, $W_2(H)$ a flattened disk with a hole, and $W_1(H)$ a complex solid with a hole and a toroidal void ($b > a$, $c > b+a$); its cross-section is shown in Figure 3.5(b). By moving the center of a sphere (a circle in the cross-section) of radius h on the boundaries of $W_1(H)$, we can get total workspace $W_1(P)$ and primary workspace $W_1^p(P)$; for small h, these have topologies similar to that of $W_1(H)$ and therefore are not shown in the figure. Assuming that $b > a+h$ [i.e., toroidal void in $W_1(P)$] and $c > b+a+h$

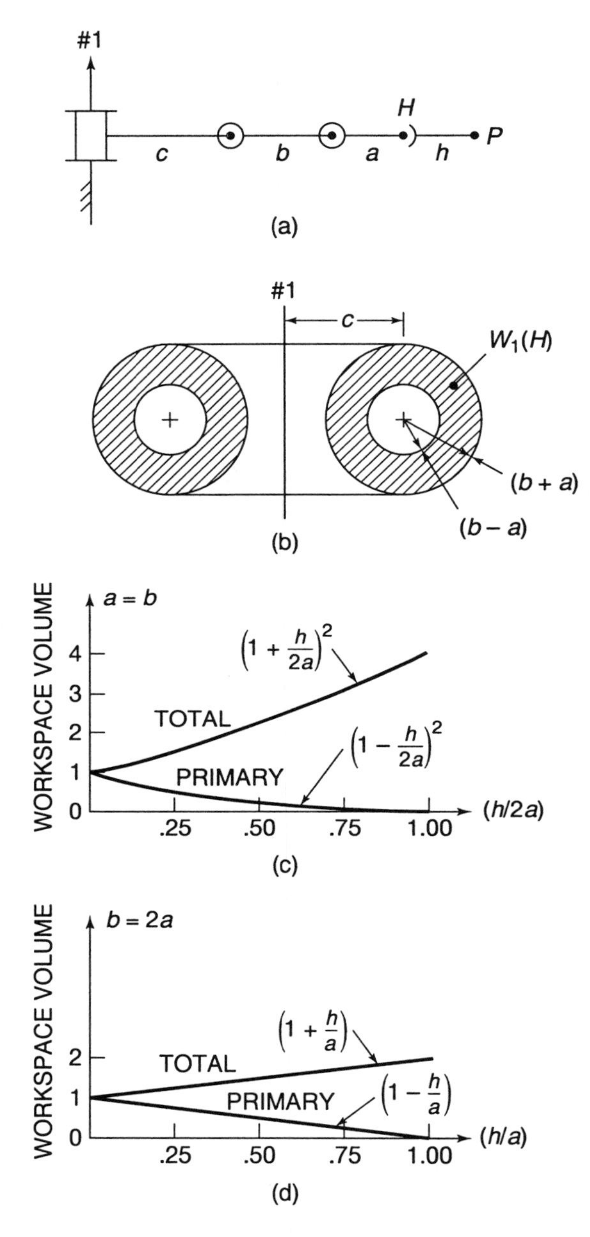

FIGURE 3.5. (a) Kinematic sketch for manipulator in Example 3.1. (b) Workspace $W_1(H)$—not to scale. (c) Plots of normalized total and primary workspace volumes with respect to the hand size h. In this case $a = b$ and $W_1(H)$ has no void. (d) Plots of normalized total and primary workspace volumes with respect to the hand size h. In this case $b = 2a$ and $W_1(H)$ has a void.

[i.e., hole in $W_1(P)$], the workspace volumes are

$$\begin{aligned} V_{\text{total}} &= (2\pi c)\cdot\pi[(a+b+h)^2-(b-a-h)^2] = 8\pi^2 bc(a+h), \\ V_{\text{primary}} &= (2\pi c)\cdot\pi[(a+b-h)^2-(b-a+h)^2] = 8\pi^2 bc(a-h). \end{aligned}$$

If $b = a$ and $c > 2a$, then the workspace $W_1(H)$ has a hole, but not the toroidal void. The total workspace $W_1(P)$ and the primary workspace $W_1^p(P)$ have similar topologies for small h. The above formulas for volumes are not valid then, and the following formulas must be used:

$$\begin{aligned} V_{\text{total}} &= (2\pi c)\cdot\pi(2a+h)^2, \\ V_{\text{primary}} &= (2\pi c)\cdot\pi(2a-h)^2. \end{aligned}$$

Figures 3.5(c) and (d) show the plots of workspace volumes with respect to the ratio (h/a) for cases when $b = a$ and $b = 2a$, respectively.

It is observed that as the hand size h increases, the volume of the total workspace increases, but the volume of the primary workspace decreases. That is, although the total workspace becomes larger, its quality deteriorates because the primary fraction becomes smaller. A similar pattern can be found for most other manipulator structures. However, there exist some manipulator structures in which the primary workspace volume has a maximum point for a nonzero value of the hand size. We will not discuss the details of such examples here, but the idealized human arm, i.e., the arm without any joint limitations, has this property. In fact, it has been claimed that the primary workspace (or dexterity) of the human arm is optimized when the hand is holding a tool such that

$$\begin{aligned} \mathcal{D}(\text{tool tip, wrist joint}) &\approx \mathcal{D}(\text{wrist joint, elbow joint}) \\ &\approx \mathcal{D}(\text{elbow joint, shoulder joint}), \end{aligned}$$

where the symbol $\mathcal{D}$ has been used for the distance operator. This property leads to some interesting anthropological speculations.

3.5 Determination of Workspace

Direct Analysis

Closed-form or iterative analyses can be used to determine points in the workspace. The workspace boundaries correspond to certain singularities of the manipulator. Analyses are carried out until the procedure breaks down. This method is very tedious, and because singularities can also occur in the interior of the workspace [see Figure 3.3(e)], it is not foolproof.

In some cases, it is possible to develop an analytical expression for the Jacobian $[J]$, and the conditions for $\det|J| = 0$ are plotted to determine the workspace boundaries.

Generation Method

The method discussed in Section 3.3 is programmed. Rotation at each joint is discretized, and displacement matrices are used to generate points of the successive workspaces, $W_k(P)$ from $W_{k+1}(P)$, etc. However, the number of points grows very rapidly with each step, and the total number generated for workspace $W_1(P)$ of 6R manipulators is in the millions and billions, even with rather coarse discretization of joint angles. For example, with 10° intervals at each revolute joint, and three data associated with the XYZ coordinates of each point, a total of $3 \times (360/10)^6 = 6.53$ billion data entries can be generated. There are two possible ways to reduce the number of stored points. First, it is noted that interior points do not affect the generation process; only the points on the interior and exterior boundary surfaces do. An interior point detection algorithm can eliminate the interior points. A simple strategy is to divide a small neighborhood of a point into eight quadrants, and if all quadrants contain points, then the point is considered an interior point and is eliminated. This does not produce a sharp definition of surface points, but rather a thin band of surface points is retained. Second, the point coordinates are rounded to two places after the decimal, and all duplicate points that result after rounding are eliminated. Using these techniques, features of workspace $W_1(P)$ can be developed by working with a few hundred points only, rather than the millions of points that would be necessary otherwise.

Radial-Slice-Layering (RSL) Method

The generation process is turned inside out. Instead of the points of workspace $W_{k+1}(P)$ being moved around the kth joint axis, the workspace $W_{k+1}(P)$ is held stationary, and a "radial" cutting plane containing the kth axis is rotated about the kth axis—a much eas-

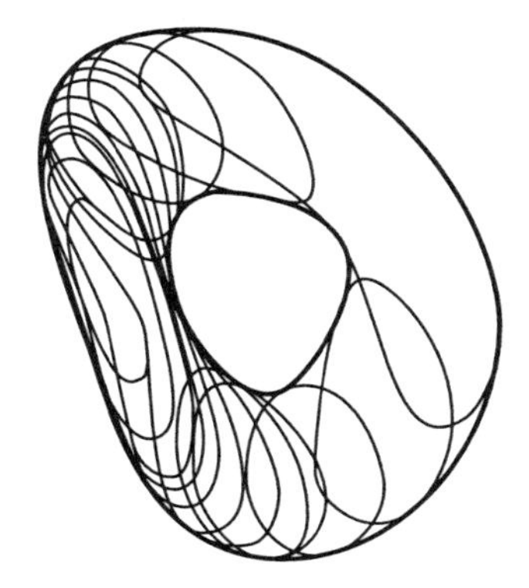

FIGURE 3.6. Cross-section of $W_k(P)$ developed by cutting a workspace surface $W_{k+1}(P)$ by multiple radial planes of axis k (i.e., all planes passing through axis k) and then collapsing these planes into a single plane. The workspace solid $W_k(P)$ will have a hole and a toroidal void.

ier task. This cutting plane produces many radial "slices" (two-dimensional) of $W_{k+1}(P)$ and this is accomplished mathematically by solving intersection problems for the cutting plane and the workspace solid. The radial "slices" are then collapsed and "layered" into a single plane. A composite two-dimensional section is developed from these "layers" and that is the cross-section of the workspace $W_k(P)$ [see Figure 3.6; the $W_{k+1}(P)$ used was a surface]. This scheme can be executed easily on CAD systems. Fine details of workspace $W_1(P)$ can be developed simply by increasing the number of radial cutting planes. The technique can be extended to include revolute joint limits as well as prismatic joints.

3.6 Free Spin of the Hand

We will now address the problem of hand spin when the hand axial direction is not collinear with the last (nth) joint axis. The cointersecting joint axis directions of a spherical equivalent wrist are represented by unit vectors $\mathbf{u}_{n-2}$, $\mathbf{u}_{n-1}$, $\mathbf{u}_n$. The hand axial direction is $\mathbf{u}_a$, and it is assumed that it passes through the wrist center H [Figure 3.7(a)]. The wrist angles are α_{n-2} (between $\mathbf{u}_{n-2}$ and $\mathbf{u}_{n-1}$) and α_{n-1} (between $\mathbf{u}_{n-1}$ and $\mathbf{u}_n$). The hand carries a tool that has a dominant axial direction $\mathbf{u}_a$, and the tool mount angle is α_n (between $\mathbf{u}_n$ and $\mathbf{u}_a$). The angle between axes $\mathbf{u}_{n-2}$ and $\mathbf{u}_a$ is variable and can be designated as α_f. Without any loss of generalization, let us assume that all α's lie in the range of $[0, \pi]$. Then, we want to investigate conditions that will ensure a 360° spin of the hand about its axial direction $\mathbf{u}_a$ when the $(n-2)$th link (i.e., the last link of the regional structure) is held fixed; the possibility of mechanical interference is ignored.

Although the derivation of these conditions is not difficult, it involves a detailed examination of certain spherical triangles, and these details will be omitted. We will focus upon the final result and its practical implications. The abovementioned free spin of the hand is possible if and only if the following two conditions are satisfied simultaneously:

$$|\alpha_f + \alpha_n - \pi| > |\alpha_{n-1} + \alpha_{n-2} - \pi| \tag{3.1}$$

and

$$|\alpha_f - \alpha_n| > |\alpha_{n-1} - \alpha_{n-2}|. \tag{3.2}$$

For the ranges of α_f where partial tool spin can occur, the existence of α_f must also be verified. It would be incorrect to conclude, after determining the full tool spin ranges, that the entire remainder of $[0°, 180°]$ will have partial tool spin.

Example 3.2. For a wrist with parameters $\alpha_{n-2} = 75°$, $\alpha_{n-1} = 80°$, and $\alpha_n = 30°$, find the ranges of angle α_f in which the tool can have full spin.

Solution: From conditions (3.1) and (3.2),

$$|\alpha_f - 150°| > 25° \tag{3.3a}$$

and

$$|\alpha_f - 30°| > 5°. \tag{3.4a}$$

An easy geometric interpretation for each condition can be obtained by marking the permissible segments for α_f in a one-dimensional range representing $[0°, 180°]$. This is better than purely algebraic manipulations of the inequalities. These inequalities lead to

$$0° < \alpha_f < 125° \quad \text{or} \quad 175° < \alpha_f < 180° \tag{3.3b}$$

and

$$0° < \alpha_f < 25° \quad \text{or} \quad 35° < \alpha_f < 180°. \tag{3.4b}$$

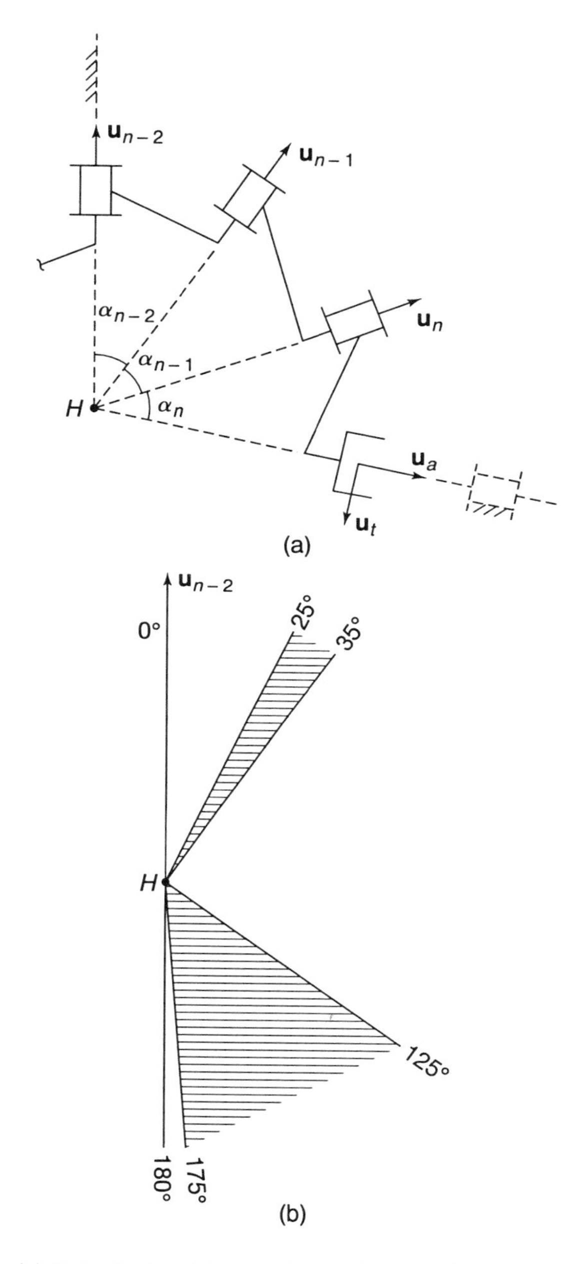

FIGURE 3.7. (a) Spherical wrist equivalent with a tool mount. The tool spin is to be about the tool axis $\mathbf{u}_a$. (b) Zones of tool axis ($\mathbf{u}_a$) orientation in which the tool can spin completely (blank zones) and partially (hatched zones). All orientations of tool axis are possible in this example.

Inequalities (3.3b) and (3.4b) hold simultaneously when

$$0^\circ < \alpha_f < 25^\circ \text{ or } 35^\circ < \alpha_f < 125^\circ \text{ or } 175^\circ < \alpha_f < 180^\circ. \qquad (3.5)$$

These are the zones in which 360° spin of the hand about its axial direction $\mathbf{u}_a$ passing through the wrist center H is possible [Figure 3.7(b)].

Example 3.3. For an orthogonal wrist, $\alpha_{n-2} = 90^\circ$, $\alpha_{n-1} = 90^\circ$, and the tool mount angle α_n is not specified. Find the ranges of α_f in which complete tool spin can occur.

Solution: Applying conditions (3.1) and (3.2),

$$|\alpha_f + \alpha_n - 180^\circ| > 0^\circ \qquad (3.6a)$$

and

$$|\alpha_f - \alpha_n| > 0^\circ. \qquad (3.7a)$$

These lead to

$$\alpha_f \neq 180^\circ - \alpha_n \qquad (3.6b)$$

and

$$\alpha_f \neq \alpha_n. \qquad (3.7b)$$

Thus, for any value of α_n, full spin of the hand is possible in the whole range of α_f, i.e., $[0, \pi]$, except when $\alpha_f = 180^\circ - \alpha_n$ or $\alpha_f = \alpha_n$. These exceptions arise because in (3.1) and (3.2), strict "$>$" inequality is used to avoid wrist singularities. Under a somewhat looser interpretation, "$\geq$" can be used in (3.1) and (3.2), and then the two exceptional cases of this example would disappear. Thus, in the case of the orthogonal wrist, all orientations of the hand axis $\mathbf{u}_a$ (which must pass through the wrist center H) are possible, and the hand can spin freely about its axis when this axis is placed in any orientation with respect to the wrist center H, ignoring the possibility of mechanical interference among the links.

An alternate solution to this problem is to provide an extra spin freedom for the tool with respect to its axis $\mathbf{u}_a$, if the extra cost for doing this can be justified.

3.7 Problems

1. Let the workspace $W_{k+1}(P)$ be a circle. Keeping axis k in the plane of this circle, sketch three distinct cases for the next generated workspace $W_k(P)$.

2. Let the solid workspace $W_{k+1}(P)$ be a donut. Keeping axis k parallel to the axis of the donut, sketch three distinct cases for the next generated solid workspace $W_k(P)$.

3. Assume that the prismatic joint (#3) in the Stanford Arm manipulator in Section 2.3 has the range from $1'$ to $5'$, when measured from the shoulder, and the shoulder offset is $0.5'$. Determine the workspace of the wrist center $W_1(H)$. How would the workspace change if the prismatic joint range could become $0'$ to $5'$?

4. For the robot configuration shown in its ZRP position in Figure P2.1, sketch the workspaces $W_3(H)$, $W_2(H)$, and $W_1(H)$ for the wrist center H.

5. For a PUMA robot, $a = 8''$, $b = c = 17''$, and $h = 4''$ (Section 2.6).

   (a) Find the cross-sections of the following: the workspace $W_1(H)$ of the wrist center; the total workspace $W_1(P)$; and the primary workspace $W_1^{\text{primary}}(P)$.

   (b) Calculate the volumes of the total and primary workspaces as well as the primary fraction. Ignore joint limits, mechanical interferences, and corner effects in workspace geometry. *Hint*: These volumes can be found without the Pappas theorem.

6. In a spherical wrist, the angles $\alpha_{n-2} = 100°$ and $\alpha_{n-1} = 50°$. The tool holding angle is $\alpha_n = 25°$. Find the ranges of the tool axis angle (α_f) for which the tool can spin completely about its own axis.

7. A robot wrist has the following specifications: $\alpha_{n-2} = 100°$ (first wrist angle), $\alpha_{n-1} = 85°$ (second wrist angle), $\alpha_n = 80°$ (tool-mount angle). Determine the ranges of tool-axis orientation angle α_f in which complete spin of the tool about its axis can occur.

8. In a spherical wrist, the wrist angles are $\alpha_{n-2} = 120°$, $\alpha_{n-1} = 60°$ and the tool mount angle $\alpha_n = 15°$; the tool axis $\mathbf{u}_a$ also passes through the spherical wrist center. Find the ranges of the angle α_f, between the axes $\mathbf{u}_{n-2}$ and $\mathbf{u}_a$, in which a 360° spin of the tool about its axial direction $\mathbf{u}_a$ is possible.

4
Dynamics and Control

4.1 Background

Robot dynamics and control is a vast subject, and separate books can be written on related topics. The objectives here are more limited. First, a brief description of dynamics and control, as they relate to robotics applications, will be provided. Main ideas and results will be presented without complete explanations and derivations. The intention is to give the reader an impression of how everything comes together in the end in a robotic system. The second objective is to show how the classical concepts of dynamics can be incorporated into the Zero Reference Position (ZRP) representation. The additional data on masses and inertias will be included in the Extended Zero Reference Position. Finally, it is hoped that the information provided here will motivate the reader to study the subjects of advanced dynamics and control in depth.

4.2 Kinematics

Kinematics is the study of motion itself. Let a rigid body have an angular velocity $\boldsymbol{\omega}$ and angular acceleration $\boldsymbol{\alpha}$. Then, the velocities and accelerations of two points A and B on the body are related as follows ($\mathbf{r}_{B/A} = \mathbf{AB}$):

$$\mathbf{V}_B = \mathbf{V}_A + \mathbf{V}_{B/A} = \mathbf{V}_A + \boldsymbol{\omega} \times \mathbf{r}_{B/A}, \tag{4.1}$$

$$\mathbf{A}_B = \mathbf{A}_A + \mathbf{A}_{B/A} = \mathbf{A}_A + \boldsymbol{\omega} \times (\boldsymbol{\omega} \times \mathbf{r}_{B/A}) + \boldsymbol{\alpha} \times \mathbf{r}_{B/A}. \tag{4.2}$$

The second and third terms in Eq. (4.2) are the normal and tangential components of acceleration $\mathbf{A}_{B/A}$. Let P be a point that can move with respect to the rigid body with a relative velocity $\mathbf{V}_{\text{rel}}$ and relative acceleration $\mathbf{A}_{\text{rel}}$, and let P' be a body point that is instantaneously coincident with point P. Then,

$$\mathbf{V}_P = \mathbf{V}_{P'} + \mathbf{V}_{\text{rel}}, \tag{4.3}$$

$$\mathbf{A}_P = \mathbf{A}_{P'} + \mathbf{A}_{\text{rel}} + \mathbf{A}_{\text{Coriolis}}; \quad \mathbf{A}_{\text{Coriolis}} = 2\boldsymbol{\omega} \times \mathbf{V}_{\text{rel}}. \tag{4.4}$$

Note the difference in the forms of Eqs. (4.3) and (4.4). The Coriolis acceleration can also explain important natural phenomena that occur due to Earth's rotation; for example, deflections of high-speed projectiles (right shift in the northern hemisphere) and swirl directions of hurricanes (counterclockwise in the northern hemisphere). These equations can be used to analyze the velocities and accelerations in multibody systems; this was done in a previous section [Eqs. (2.49) to (2.52)] to derive the manipulator Jacobian.

4.3 Kinetics

Kinetics is concerned with the cause and effect interrelationships among forces and motion. In a forward or direct dynamics problem, the forces acting on a system are given, and we must find the resulting motion. It is a difficult problem to solve. In an inverse dynamics problem, the motion of the system is given, and we must find the forces necessary to sustain the motion. Inverse dynamics problems are much simpler to solve than the forward/direct dynamics problems.

Particles

For a single particle, Newton's second law states that

$$\mathbf{F} = m\mathbf{a}. \tag{4.5}$$

For a system of particles with mass center at G,

$$\mathbf{F}^{\text{ext}} = d\mathbf{L}/dt = m\mathbf{a}_G, \tag{4.6}$$

$$\mathbf{M}_G^{\text{ext}} = d\mathbf{H}_G/dt, \tag{4.7}$$

$$\mathbf{L} = \sum_i m_i \mathbf{v}_i; \quad \mathbf{H}_G = \sum_i \mathbf{r}_{i/G} \times m_i \mathbf{v}_i, \tag{4.8}$$

where $\mathbf{L}$ is linear momentum, $\mathbf{H}_G$ is angular momentum with respect to mass center G, $\mathbf{F}^{\text{ext}}$ and $\mathbf{M}_G^{\text{ext}}$ are the resultants of external forces and moments (about G), and $\mathbf{r}_{i/G}$ is the position of particle i with respect to the system mass center G. The force equation (4.6) is practically useful, but the moment equation (4.7) is not useful because the computation of $\mathbf{H}_G$ is difficult. Also note that only external forces and moments are used because all of the internal forces cancel out due to the action–reaction principle.

Rigid Bodies

In a planar motion, all points in the body move in parallel planes. There is no restriction on the physical size of the body. For the planar motion of a rigid body, when the xy-plane of motion (i) passes through the center of mass G, and (ii) is also a plane of symmetry, Eqs. (4.6) and (4.7) become

$$F_x^{\text{ext}} = ma_{Gx}; \quad F_y^{\text{ext}} = ma_{Gy}; \quad M_G^{\text{ext}} = I_G \alpha, \tag{4.9}$$

where I_G is the moment of inertia about an axis through G and parallel to the z axis. If k is the radius of gyration, then $I_G = mk^2$.

The rotational resistance of a rigid body in three dimensions is measured by the inertia tensor $[I]$ or $[I_G]$.

$$[I] = \begin{bmatrix} I_{xx} & I_{xy} & I_{xz} \\ I_{xy} & I_{yy} & I_{yz} \\ I_{xz} & I_{yz} & I_{zz} \end{bmatrix}, \tag{4.10}$$

where

$$I_{xx} = \int_m (y^2 + z^2) dm, \ldots, I_{xy} = I_{yx} = -\int_m xy\, dm, \ldots. \tag{4.11}$$

I_{xx}, I_{yy}, I_{zz} are called moments of inertia with respect to axes x, y, z, respectively; I_{xy}, I_{yz}, I_{xz} are called products of inertia. Note the negative signs in the definitions of the products of inertia; they are necessary to have a proper form for the inertia tensor in Eq. (4.10).

If the xyz system located at G (i.e., a centroidal system) is not a body system, then the inertias are time dependent [but this variation can be accounted for, see Eq. (4.14)]; if xyz is a centroidal body system (i.e., located at G), then the inertias are a set of constants [Figure 4.1(a)].

If there is a noncentroidal body system $x'y'z'$ located at point A such that the centroidal xyz system is translated by (a, b, c) with respect to it, i.e., $x' = x + a$, $y' = y + b$, $z' = z + c$ [Figure 4.1(b)], then the moments of inertia are related by the parallel axis theorem as follows:

$$[I_A] = [I_G] + m \begin{bmatrix} b^2 + c^2 & -ab & -ac \\ -ab & c^2 + a^2 & -bc \\ -ac & -bc & a^2 + b^2 \end{bmatrix}. \tag{4.12}$$

Equation (4.12) is valid for transfers between centroidal and noncentroidal (translated or parallel) body systems. To relate $[I_A]$ and $[I_B]$ for two parallel noncentroidal systems, we must go through $[I_G]$ for the parallel centroidal system.

If two body systems $x''y''z''$ and xyz have common origins but differ in orientation such that the rotation matrix $[R]$ will rotate the xyz system from its coincident position with the $x''y''z''$ system to its actual orientation (i.e., the matrix $[R]$ will change coordinates from the xyz system to the $x''y''z''$ system), then inertia matrices $[I'']$ and $[I]$ are related as

$$[I''] = [R][I][R]^t; \quad [I] = [R]^t[I''][R]. \tag{4.13}$$

If the $x''y''z''$ system is not a body system but, say, a translated base system with the same origin as the xyz body system, then the rotation matrix R becomes time dependent, i.e., $R = R(t)$, and the corresponding time-dependent inertia matrix $[I''(t)]$ is

$$[I''(t)] = [R(t)][I][R(t)]^t; \quad [I] = \text{constant inertia matrix}. \tag{4.14}$$

In general, there exists a unique body system in which the products of inertia terms vanish (i.e., $I_{xy} = I_{yz} = I_{xz} = 0$) and the inertia matrix becomes diagonal. This is called the principal body system (1–2–3), and the three inertias are called principal inertias (I_1, I_2, I_3); these are, respectively, the eigenvectors and eigenvalues

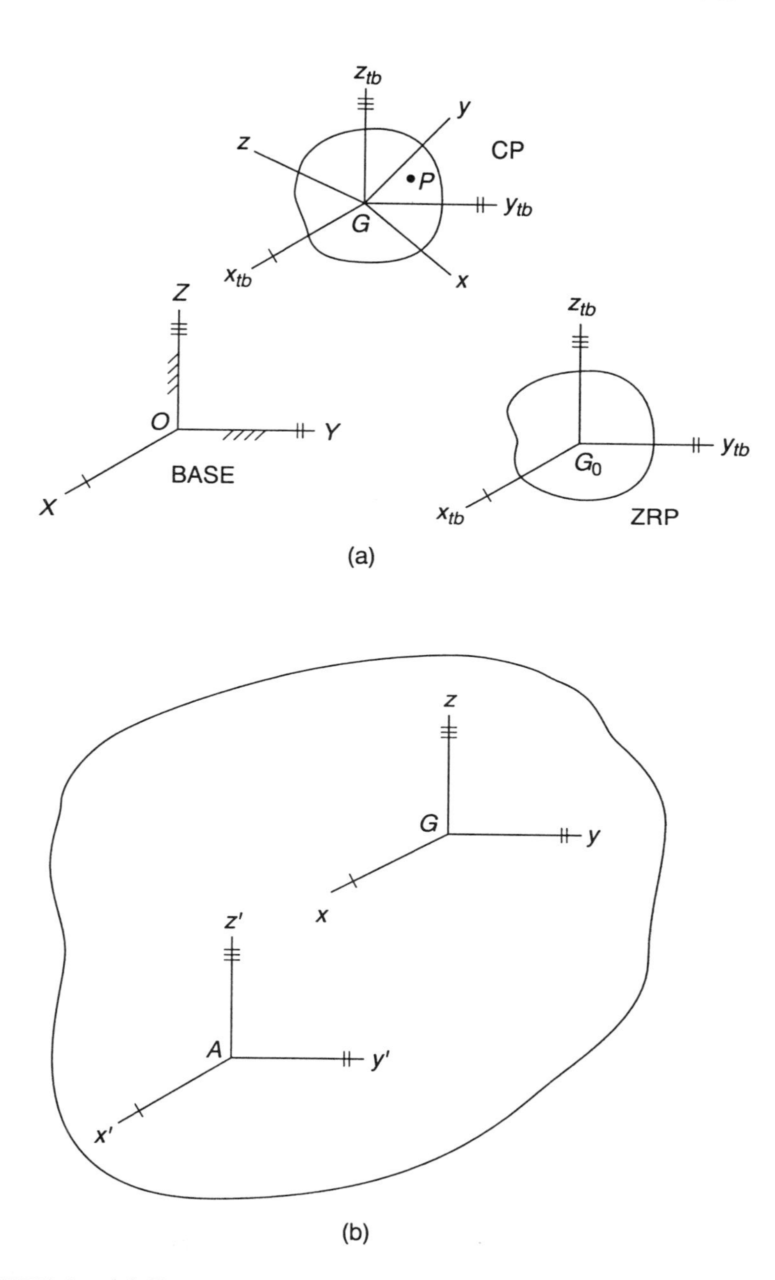

FIGURE 4.1. (a) Shown are the translated base system at ZRP and the current position (CP). Also shown is the body system xyz at CP. The mass center is at G. (b) Centroidal body system xyz and a parallel (axis-wise) noncentroidal body system $x'y'z'$.

of the inertia matrix. If the body has a plane of symmetry, then the normal to the plane is one of the principal axes (e.g., parallelepiped, cylinder, sphere); if the body has an axis of symmetry, then that axis is also one of the principal axes (e.g., parallelepiped, cylinder, sphere). Exceptions to uniqueness of the principal axis system occur when the body has one (e.g., cylinder) or more (e.g., sphere) axes of symmetry.

It was mentioned previously that the moment equation (4.7) is not particularly useful for general systems of particles, but it can be modified for rigid bodies. The angular momentum for a rigid body in a centroidal body system xyz can be simplified with the velocity equation for any point P [Figure 4.1(a)] as

$$\mathbf{v}_P = \mathbf{v}_G + \boldsymbol{\omega} \times \mathbf{r}_{P/G}; \quad \text{or, } \mathbf{v} = \mathbf{v}_G + \boldsymbol{\omega} \times \mathbf{r}, \tag{4.15}$$

$$\mathbf{H}_G = \int_m \mathbf{r} \times (\mathbf{v}\, dm) = \left(\int_m \mathbf{r}\, dm\right) \times \mathbf{v}_G - \int_m \mathbf{r} \times (\mathbf{r} \times \boldsymbol{\omega})\, dm. \tag{4.16}$$

The first integral leads to $m\mathbf{r}_G = 0$ because the origin of the xyz body system is at G. To simplify the second integral, we switch from the vector notation to the matrix notation. The Cartesian vectors become column vectors, and the cross-product operation $\mathbf{r}\,\times$ becomes premultiplication by the skew-symmetric matrix $[r]$, where

$$[r] = \begin{bmatrix} 0 & -z & y \\ z & 0 & -x \\ -y & x & 0 \end{bmatrix}. \tag{4.17}$$

Then, $\mathbf{r} \times (\mathbf{r} \times \boldsymbol{\omega})$ is replaced by $[r]^2\boldsymbol{\omega}$, and we get

$$\mathbf{H}_G = [I_G]\boldsymbol{\omega}; \quad [I_G] = -\int_m [r]^2\, dm. \tag{4.18}$$

The components of $\mathbf{H}_G$ and the elements of $[I_G]$ [also see Eqs. (4.10) and (4.11)] are found in the centroidal body system xyz.

Generalized Euler's Equation (Body System Form)

In Eq. (4.7), the derivative with respect to a fixed frame XYZ (base) is implied, but $\mathbf{H}_G$ in Eq. (4.18) has components in the body system (xyz).

$$\mathbf{M}_G^{\text{ext}} = (d\mathbf{H}_G/dt)_{XYZ}, \tag{4.7}$$

$$= (d\mathbf{H}_G/dt)_{xyx} + [\Omega](\mathbf{H}_G)_{xyz}. \tag{4.19}$$

Equation (4.19) accounts for the fact that the body system xyz with unit vectors **ijk** is itself moving with angular velocity $\boldsymbol{\omega}$ and vectorially,

$$(d\mathbf{i}/dt) = \boldsymbol{\omega} \times \mathbf{i}, \quad (d\mathbf{j}/dt) = \boldsymbol{\omega} \times \mathbf{j}, \quad (d\mathbf{k}/dt) = \boldsymbol{\omega} \times \mathbf{k},$$

and in matrix notation,

$$(d\mathbf{i}/dt) = [\Omega]\mathbf{i}, \quad (d\mathbf{j}/dt) = [\Omega]\mathbf{j}, \quad (d\mathbf{k}/dt) = [\Omega]\mathbf{k}.$$

The definition of skew-symmetric angular velocity matrix $[\Omega]$ [see Eq. (2.23)] is

$$[\Omega] = \begin{bmatrix} 0 & -\omega_z & \omega_y \\ \omega_z & 0 & -\omega_x \\ -\omega_y & \omega_x & 0 \end{bmatrix}.$$

Finally, substituting Eq. (4.18),

$$\mathbf{M}_G^{\text{ext}} = [I_G]\boldsymbol{\alpha} + [\Omega][I_G]\boldsymbol{\omega}. \tag{4.20}$$

Equation (4.20) is the generalized Euler's equation for the centroidal body system (xyz). Equations (4.6) and (4.20) are the equations of motion for spatial motion of a rigid body. These can be written in D'Alembert's form, using centroidal body system xyz, as

$$\mathbf{F}^{\text{ext}} + \mathbf{F}^{\text{inertia}} = \mathbf{0}; \quad \mathbf{F}^{\text{inertia}} = -m\mathbf{a}_G, \tag{4.21}$$

$$\mathbf{M}_G^{\text{ext}} + \mathbf{M}_G^{\text{inertia}} = \mathbf{0}; \quad \mathbf{M}_G^{\text{inertia}} = -[I_G]\boldsymbol{\alpha} - [\Omega][I_G]\boldsymbol{\omega}. \tag{4.22}$$

In D'Alembert's approach, the inertia effects have been brought into the "active" force domain, and the resulting equations are called the equations of dynamic equilibrium.

In the principal centroidal body system, the inertia moments become

$$\begin{aligned} M_{G1}^{\text{inertia}} &= -I_1\alpha_1 - (I_3 - I_2)\omega_2\omega_3, \\ M_{G2}^{\text{inertia}} &= -I_2\alpha_2 - (I_1 - I_3)\omega_1\omega_3, \\ M_{G3}^{\text{inertia}} &= -I_3\alpha_3 - (I_2 - I_1)\omega_1\omega_2. \end{aligned} \tag{4.23}$$

Generalized Euler's Equations (Non-Body System Form)

An alternate form of the moment equation (4.20) can be derived for the translated base (tb) system $x_{\mathrm{tb}}y_{\mathrm{tb}}z_{\mathrm{tb}}$ located at G [Figure 4.1(a)]. The translated base system, although located at the mass center G of the moving body, remains axis-wise parallel to the fixed base system XYZ. The centroidal body system is xyz. Let $[R]$ be the matrix that changes coordinates from the body system to the translated body system. Then, all of the quantities can be referred to in the translated base system as follows; when a coordinate system is not indicated, the centroidal body system is implied.

$$(\boldsymbol{\omega})_{\mathrm{tb}} = [R]\boldsymbol{\omega}, \quad (\boldsymbol{\alpha})_{\mathrm{tb}} = [R]\boldsymbol{\alpha}, \quad (\mathbf{M}_G^{\mathrm{ext}})_{\mathrm{tb}} = [R]\mathbf{M}_G^{\mathrm{ext}}, \tag{4.24}$$

$$[\Omega]_{\mathrm{tb}} = [R][\Omega][R]^t, \quad [I_G]_{\mathrm{tb}} = [R][I_G][R]^t. \tag{4.25}$$

The transformations of 3×1 column vector entities are given by Eq. (4.24), and those of 3×3 matrix quantities are given by Eq. (4.25), the latter being similarity transformations.

From the above transformations and Eq. (4.20),

$$\begin{aligned}(\mathbf{M}_G^{\mathrm{ext}})_{\mathrm{tb}} &= [R]\mathbf{M}_G^{\mathrm{ext}} \\ &= [R]\{[I_G]\boldsymbol{\alpha} + [\Omega][I_G]\boldsymbol{\omega}\} \\ &= [R][I_G][R]^t\,[R]\boldsymbol{\alpha} + [R][\Omega][R]^t\,[R][I_G][R]^t\,[R]\boldsymbol{\omega}.\end{aligned}$$

Note that $[R]^t[R] = I$ (identity matrix) has been used to add terms in three places. The moment equation with respect to the translated base system $(x_{\mathrm{tb}}y_{\mathrm{tb}}z_{\mathrm{tb}})$ at G is then

$$(\mathbf{M}_G^{\mathrm{ext}})_{\mathrm{tb}} = [I_G]_{\mathrm{tb}}(\boldsymbol{\alpha})_{\mathrm{tb}} + [\Omega]_{\mathrm{tb}}[I_G]_{\mathrm{tb}}(\boldsymbol{\omega})_{\mathrm{tb}}. \tag{4.26}$$

This is the generalized Euler's equation for a centroidal nonbody system, particularly the translated base system located at the mass center G. Equations (4.20) and (4.26) have similar appearances but are fundamentally different. In Eq. (4.20), the centroidal body system xyz is used, and the inertia properties of the body are constants. In Eq. (4.26), the translated base system $(x_{\mathrm{tb}}y_{\mathrm{tb}}z_{\mathrm{tb}})$ at G is used, and the inertia matrix $[I_G]_{\mathrm{tb}} = [R(t)][I_G][R(t)]^t$ is time dependent. But the fact that Eqs. (4.20) and (4.26) have similar forms is remarkable, and this can be argued in hindsight as follows: imagine a special body system that is coincident with the translated base system instantaneously, and then the same form of equation should be valid for both, but the quantities must be referred to the coordinate system appropriately. However, such an intuitive argument cannot substitute for proof, which has been provided above, because the form in Eq. (4.26) is not commonly found in the literature, and most elementary and intermediate level books on dynamics repeatedly warn the reader that only a body system must be used, or else something terrible may happen. Here we have seen the precise effect of using a nonbody system. Equations (4.24)–(4.26) are also the equations that will allow us to incorporate the classical concepts of dynamics (Newton–Euler equations, D'Alembert's principle, and Lagrangian mechanics) into the framework of the Extended Zero Reference Position Approach.

Generalized Euler's Equation (ZRP Form)

In the extended ZRP approach, the additional inertia data are given with respect to the translated base system located at link mass center G in the zero reference position of the link [Figure 4.1(a)], i.e., constant matrix $[I_{G0}]$ or $[I_{G0}]_{\text{tb}}$. The time-dependent link rotation matrix that orients the link from ZRP to the current (CP) position is $[R(t)]$. Then, the equations of motion for the link in the current position, and referred to in the translated base system at G, are given by Eqs. (4.6) and (4.26):

$$(\mathbf{F}^{\text{ext}})_{\text{tb}} = m(\mathbf{a}_G)_{\text{tb}}, \tag{4.6}$$

$$(\mathbf{M}_G^{\text{ext}})_{\text{tb}} = [I_G(t)]_{\text{tb}}(\boldsymbol{\alpha})_{\text{tb}} + [\Omega]_{\text{tb}}[I_G(t)]_{\text{tb}}(\boldsymbol{\omega})_{\text{tb}}. \tag{4.26}$$

The time-dependent inertia matrix now is

$$[I_G(t)]_{\text{tb}} = [R(t)][I_{G0}]_{\text{tb}}[R(t)]^t.$$

It would be desirable to reformulate these equations such that the time dependence of the inertia matrix is eliminated. This can be done by defining "*" superscripted quantities by premultiplying all of the 3×1 vector quantities for the link by the transpose of the link rotation matrix, i.e., $[R(t)]^t$, and applying reverse similarity transformations to 3×3 quantities.

$$(\boldsymbol{\omega})^* = [R]^t(\boldsymbol{\omega})_{\text{tb}}, \quad (\boldsymbol{\alpha})^* = [R]^t(\boldsymbol{\alpha})_{\text{tb}}, \quad (\mathbf{a}_G)^* = [R]^t(\mathbf{a}_G)_{\text{tb}},$$

$$(\mathbf{F}^{\text{ext}})^* = [R]^t(\mathbf{F}^{\text{ext}})_{\text{tb}}, \quad (\mathbf{M}_G^{\text{ext}})^* = [R]^t(\mathbf{M}_G^{\text{ext}})_{\text{tb}}, \tag{4.27}$$

$$[\Omega]^* = [R]^t[\Omega]_{\text{tb}}[R], \quad [I_G]^*_{\text{tb}} = [R]^t[I_G]_{\text{tb}}[R] = [I_{G0}]_{\text{tb}}.$$

Caution must be exercised in dealing with multibody systems because each link has its own link rotation matrix. However, just as regular quantities (velocities, accelerations, forces, moments) for adjacent links can be related, so can the corresponding "$*$" quantities for adjacent links. Thus, the time-consuming transformations among the regular and "$*$" quantities, as seen in the definitions (4.27), are avoided.

The modified equations of motion, derived from Eqs. (4.6) and (4.26), are the Newton–Euler equations for the ZRP approach:

$$(\mathbf{F}^{\text{ext}})^* = m(\mathbf{a}_G)^*, \tag{4.28}$$

$$(\mathbf{M}_G^{\text{ext}})^* = [I_{G0}](\boldsymbol{\alpha})^* + [\Omega]^*[I_{G0}](\boldsymbol{\omega})^*. \tag{4.29}$$

The advantages of the "$*$" notation are that

(i) the inertia matrix $[I_{G0}]$ is a constant matrix,

(ii) the body vectors ($\mathbf{c}$) of the link are referred back to the ZRP position ($\mathbf{c}_o$)—this follows from the definition of current orientation of a body vector:

$$\mathbf{c}(t) = [R(t)]\mathbf{c}_o, \quad \mathbf{c}^* = [R(t)]^t\mathbf{c}(t) = \mathbf{c}_o,$$

(iii) the same link rotation matrix $[R(t)]$ that is available from the ZRP kinematic analysis can be used for dynamic analysis,

(iv) the amount of computations is reduced significantly,

(v) closed-form equations of motions can be derived for many industrial manipulators by using symbolic manipulation programs.

In more conventional terminology, the "$*$" equations correspond to the case when the translated base system at G in the ZRP position is also chosen as a special centroidal body system, and then the $(..)^*$ and $[..]^*$ quantities are essentially expressed in this special body system when the body is in its current position.

The ZRP approach gives us more flexibility in formulating the kinematics of manipulators. However, if one is using the ZRP approach and wants to get results identical to those obtained with the

Denavit–Hartenberg approach in conjunction with classical Newton–Euler equations, then one must do the following: choose the ZRP position as the zero position of the DH-notation, and in the conventional kinetic analysis, choose the centroidal link body systems as the translated base systems at link CGs in this zero DH position.

4.4 Recursion Relations

In a serially connected robot, kinematic and kinetic relations can be obtained by forward (base-to-hand) and backward (hand-to-base) recursions. The notation used in this section will be a mixture of vector and matrix notations. Let $\mathbf{a}$ and $\mathbf{b}$ be either the Cartesian vectors (with $\mathbf{ijk}$ components) or column vectors (3×1 matrix representation). The dot product can be represented as $\mathbf{a} \cdot \mathbf{b}$ (Cartesian) or $\mathbf{a}^t\mathbf{b}$ (matrix). The cross product can be represented as $\mathbf{c} = \mathbf{a} \times \mathbf{b}$ (Cartesian) or $\mathbf{c} = [A]\mathbf{b}$ (matrix), where $[A]$ is the skew-symmetric form of $\mathbf{a}$. Although not common in pure mathematics literature, we have already treated $\mathbf{a} \times \mathbf{b}$ as a 3×1 column vector when this notation was used for certain elements of manipulator Jacobian. In this section, we will go further and mix some matrix and vector operations, and the purpose of this discussion is to clarify the intent. This will happen when we apply a rotation operator (or coordinate transformation) to a formal vector equation. It is possible to avoid this mixing if we use dyadics or tensors, but that will not be done here.

It is important to remember that the vector dot and cross products have the physical meaning of projection and enclosed area, respectively, and if rigid-body rotation or coordinate transformation is applied to the vectors, either the physical meaning is not changed, for example, the dot product,

$$\mathbf{a} \cdot \mathbf{b} = R\mathbf{a} \cdot R\mathbf{b} \quad \text{or} \quad \mathbf{a}^t\mathbf{b} = (R\mathbf{a})^t(R\mathbf{b}),$$

or it is changed in a consistent manner, for example the cross product,

$$R\mathbf{c} = R\mathbf{a} \times R\mathbf{b} \quad \text{or} \quad R\mathbf{c} = R[A]\mathbf{b} = R[A]R^tR\mathbf{b}.$$

Purists will surely object to expressions such as $R\mathbf{a} \cdot R\mathbf{b}$ and $R\mathbf{a} \times R\mathbf{b}$, but they have better appeal than $(R\mathbf{a})^t(R\mathbf{b})$ and $R[A]R^tR\mathbf{b}$, respectively. The experience has been that most students will evaluate

$R\mathbf{a} \cdot R\mathbf{b}$ and $R\mathbf{a} \times R\mathbf{b}$ as intended, i.e., find the vectors $R\mathbf{a}$ and $R\mathbf{b}$ first, and then find their dot or cross product, and express the final result in Cartesian vector or matrix form. A simple rule to remember is that when a rigid-body rotation or transformation is applied to a formal vector expression, it must be applied to every vector in the expression. For example, for the volume contained by vectors $\mathbf{a}$, $\mathbf{b}$, and $\mathbf{c}$, the triple product

$$\mathbf{a} \cdot (\mathbf{b} \times \mathbf{c}) = R\mathbf{a} \cdot (R\mathbf{b} \times R\mathbf{c}),$$

and the vector identity for triple cross product

$$\mathbf{a} \times (\mathbf{b} \times \mathbf{c}) = (\mathbf{a} \cdot \mathbf{c})\mathbf{b} - (\mathbf{a} \cdot \mathbf{b})\mathbf{c}$$

leads to

$$\begin{aligned} [R]\{\mathbf{a} \times (\mathbf{b} \times \mathbf{c})\} &= (R\mathbf{a} \cdot R\mathbf{c})[R]\mathbf{b} - (R\mathbf{a} \cdot R\mathbf{b})[R]\mathbf{c} \\ &= (\mathbf{a} \cdot \mathbf{c})[R]\mathbf{b} - (\mathbf{a} \cdot \mathbf{b})[R]\mathbf{c}. \end{aligned}$$

The brackets [..] are used for matrices when they are helpful, but they are dropped when they would only lead to cumbersome writing. Things become intuitively simple if the vector operations are brought into the matrix domain by treating them as black-box operators but expressing the final result in the matrix form. In multibody problems, the kinematic relations for velocities and accelerations have their best forms in the Cartesian vector notation, while the rigid-body rotations or transformations, and generalized Euler's equations have their best forms in the matrix notation.

The link rotation matrix for the kth link (which exists between axes $k-1$ and k) is

$$R_k^L = R(\theta_1, \mathbf{u}_{10})R(\theta_2, \mathbf{u}_{20}) \ldots R(\theta_{k-1}, \mathbf{u}_{k-1,0}). \tag{4.30}$$

If a joint is prismatic, then the corresponding θ is set to zero. Relations between the link rotation matrices for adjacent links k and $k+1$ are

$$R_{k+1}^L = R_k^L R(\theta_k, \mathbf{u}_{k0}), \tag{4.31}$$

$$(R_k^L)^t = R(\theta_k, \mathbf{u}_{k0})(R_{k+1}^L)^t. \tag{4.32}$$

Consider the forward recursion for $\boldsymbol{\omega}$ and $\boldsymbol{\alpha}$ (expressed in the fixed base system) across the kth revolute joint:

$$\boldsymbol{\omega}_{k+1} = \boldsymbol{\omega}_k + (d\theta_k/dt)\mathbf{u}_k, \tag{4.33}$$

$$\boldsymbol{\alpha}_{k+1} = \boldsymbol{\alpha}_k + (d^2\theta_k/dt^2)\mathbf{u}_k + (d\theta_k/dt)\boldsymbol{\omega}_k \times \mathbf{u}_k. \tag{4.34}$$

The current orientation of the kth joint axis is $\mathbf{u}_k$. If we change to "*" quantities, which are defined by premultiplying regular 3×1 vector quantities with the transpose of the respective link rotation matrices, then the result after simplification is

$$\begin{aligned}(\boldsymbol{\omega}_{k+1})^* &= [R(\theta_k, \mathbf{u}_{k0})]^t\{(\boldsymbol{\omega}_k)^* + (d\theta_k/dt)\mathbf{u}_{k0}\} \\ &= [R(\theta_k, \mathbf{u}_{k0})]^t(\boldsymbol{\omega}_k)^* + (d\theta_k/dt)\mathbf{u}_{k0},\end{aligned} \tag{4.35}$$

$$\begin{aligned}(\boldsymbol{\alpha}_{k+1})^* &= [R(\theta_k, \mathbf{u}_{k0})]^t\{(\boldsymbol{\alpha}_k)^* + (d^2\theta_k/dt^2)\mathbf{u}_{k0} + (d\theta_k/dt)(\boldsymbol{\omega}_k)^* \times \mathbf{u}_{k0}\} \\ &= [R(\theta_k, \mathbf{u}_{k0})]^t(\boldsymbol{\alpha}_k)^* + (d^2\theta_k/dt^2)\mathbf{u}_{k0} \\ &\quad + (d\theta_k/dt)[R(\theta_k, \mathbf{u}_{k0})]^t(\boldsymbol{\omega}_k)^* \times \mathbf{u}_{k0}.\end{aligned} \tag{4.36}$$

Note that

$$\mathbf{u}_k = R_k^L \mathbf{u}_{k0} \ \text{ implies } \ (\mathbf{u}_k)^* = \mathbf{u}_{k0}.$$

Also note that $\mathbf{u}_{k0}$ is not affected by $[R(\theta_k, \mathbf{u}_{k0})]^t$. The forward recursion among the "*" quantities in Eqs. (4.35)–(4.36) is much simpler computationally than that in Eqs. (4.33)–(4.34) for regular quantities; in fact, they are simple enough to derive closed-form expressions for many industrial robots either manually or by using algebraic manipulation programs (REDUCE, MAPLE, etc.). Also note that Eqs. (4.35)–(4.36) have the mixed vector and matrix notation of the type discussed previously.

The backward recursion, by rearrangement of Eqs. (4.35) and (4.36), is as follows:

$$(\boldsymbol{\omega}_k)^* = [R(\theta_k, \mathbf{u}_{k0})](\boldsymbol{\omega}_{k+1})^* - (d\theta_k/dt)\mathbf{u}_{k0}, \tag{4.37}$$

$$(\boldsymbol{\alpha}_k)^* = [R(\theta_k, \mathbf{u}_{k0})](\boldsymbol{\alpha}_{k+1})^* - (d^2\theta_k/dt^2)\mathbf{u}_{k0} - (d\theta_k/dt)(\boldsymbol{\omega}_k)^* \times \mathbf{u}_{k0}. \tag{4.38}$$

Other expressions that follow will be given directly in terms of the "*" quantities, and the transformation of the type preceding from regular to "*" quantities will be implied. Although these transformations are not simple, we are going to skip the details to save space; also note that when the free-body diagrams are drawn for the dynamical equilibrium of links, only regular quantities must be shown in the figures, although the equations are presented in their final form involving the "*" quantities.

The corresponding relations across the kth prismatic joint are

$$(\boldsymbol{\omega}_{k+1})^* = [R(\theta_k, \mathbf{u}_{k0})]^t(\boldsymbol{\omega}_k)^*, \quad (\boldsymbol{\alpha}_{k+1})^* = [R(\theta_k, \mathbf{u}_{k0})]^t(\boldsymbol{\alpha}_k)^*, \tag{4.39}$$

$$(\boldsymbol{\omega}_k)^* = [R(\theta_k, \mathbf{u}_{k0})](\boldsymbol{\omega}_{k+1})^*, \quad (\boldsymbol{\alpha}_k)^* = [R(\theta_k, \mathbf{u}_{k0})](\boldsymbol{\alpha}_{k+1})^*. \tag{4.40}$$

However, if the kth joint is prismatic, then $[R(\theta_k, \mathbf{u}_{k0})] = [R(0, \mathbf{u}_{k0})] = I$, and we have

$$(\boldsymbol{\omega}_k)^* = (\boldsymbol{\omega}_{k+1})^*, \quad (\boldsymbol{\alpha}_k)^* = (\boldsymbol{\alpha}_{k+1})^*. \tag{4.41}$$

Similar recursions can be derived for $(\mathbf{v}_{G_k})^*$, $(\mathbf{v}_{G_{k+1}})^*$, $(\mathbf{a}_{G_k})^*$, $(\mathbf{a}_{G_{k+1}})^*$. It should be noted that while only the acceleration recursions are used in the Newton–Euler formulation, the velocity recursions will be used later for the Lagrangian formulation. The writing of these relations becomes very tedious, and their forms depend upon whether the kth joint is revolute or prismatic and, if the kth joint is prismatic, upon its mechanical construction. The specific details of prismatic joint construction enter consideration because the linear velocity and acceleration relations utilize coincident points at joint centers. The link containing the sliding collar still has the current coincident point at the prismatic joint center as the original ZRP body point. However, the current coincident point on the link containing the rod element is no longer the original ZRP body point of that link, and adjustment must be made to account for sliding at the prismatic joint.

Now consider the dynamic equilibrium of link k as shown in Figure 4.2(b). Although we will work with "$*$" quantities, it would not be correct to show these on the figure, and only the regular quantities are shown. At the link mass center G_k, the weight, the inertia force $\mathbf{F}_k^{\text{inertia}}$, and the moment $\mathbf{M}_{G_k}^{\text{inertia}}$ are shown as $\mathbf{W}_k$, $\mathbf{F}_k$, and $\mathbf{M}_k$, respectively. Point P_{k-1}^k is a body point belonging to link k (note the superscript) at the $(k-1)$th joint center; similarly point P_k^k is a body point of link k at the kth joint center [see Figure 4.2(a)]. The reaction force and moment exerted by link $(k-1)$ upon link k are $\mathbf{N}_{k-1}$ and $\mathbf{T}_{k-1}$, respectively (note the subscript); in view of this definition, $(-\mathbf{N}_k)$ and $(-\mathbf{T}_k)$ are shown at joint k. Applying D'Alembert's principle [Eqs. (4.21) and (4.22)] and summing moments about the $(k-1)$th joint center, we obtain

$$\mathbf{N}_{k-1} + \mathbf{W}_k + \mathbf{F}_k + (-\mathbf{N}_k) = 0, \tag{4.42}$$

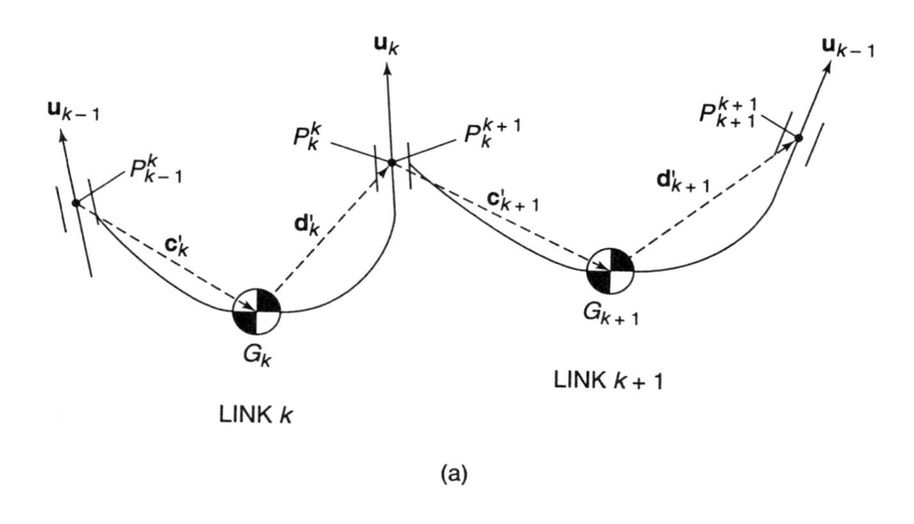

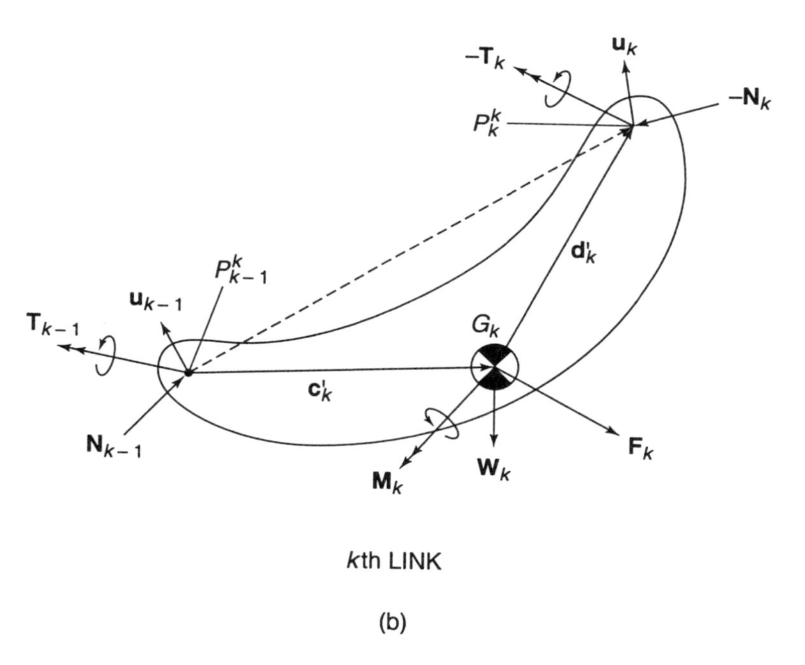

FIGURE 4.2. (a) Two adjacent links are shown. Note the definitions of the coincident points at the joint centers and of the current body vectors $\mathbf{c}'$ and $\mathbf{d}'$. (b) Dynamic equilibrium of the kth link. The joint reaction forces and moments are $(\mathbf{N}, \mathbf{T})$, the inertial force and moment are $(\mathbf{F}, \mathbf{M})$, and the weight is $\mathbf{W}$. The body vectors are $\mathbf{c}'$ and $\mathbf{d}'$.

$$\mathbf{T}_{k-1}+\mathbf{c}'_k\times(\mathbf{W}_k+\mathbf{F}_k)+\mathbf{M}_k+(\mathbf{c}'_k+\mathbf{d}'_k)\times(-\mathbf{N}_k)+(-\mathbf{T}_k)=0, \quad (4.43)$$

where

$$\mathbf{c}'_k = \mathbf{P}^k_{k-1}\mathbf{G}_k \quad \text{and} \quad \mathbf{d}'_k = \mathbf{G}_k\mathbf{P}^k_k$$

are body vectors in the current position of link k. Body vector $\mathbf{c}'_k$ goes from the $(k-1)$th joint center to the mass center of G_k link k, and body vector $\mathbf{d}'_k$ goes from the mass center G_k to the kth joint center. These body vectors in the ZRP position are $\mathbf{c}_{ko}$ and $\mathbf{d}_{ko}$, respectively. Due to link rotations, these become $\mathbf{c}_k$ and $\mathbf{d}_k$, respectively, in the current position. If both joints of link k are revolute joints, or if they are prismatic but the collar parts belong to link k, then

$$\mathbf{c}'_k = \mathbf{c}_k \quad \text{and} \quad \mathbf{d}'_k = \mathbf{d}_k.$$

However, if the rod part of the $(k-1)$th prismatic joint belongs to link k, then

$$\mathbf{c}'_k = \mathbf{c}_k - s_{k-1}\mathbf{u}_{k-1},$$

and if the rod part of the kth prismatic joint belongs to link k, then

$$\mathbf{d}'_k = \mathbf{d}_k + s_k\mathbf{u}_k.$$

Note the signs of the sliding amounts at the joints; the adjustment occurs at the tail end of vector $\mathbf{c}'_k$ and at the arrowhead of vector $\mathbf{d}'_k$.

Now change to "$*$" quantities as follows:

$$(\mathbf{F}_k)^* = (\mathbf{F}_k^{\text{inertia}})^* = -m(\mathbf{a}_{G_k})^*, \quad (4.44)$$

$$(\mathbf{M}_k)^* = (\mathbf{M}_{G_k}^{\text{inertia}})^* = -[I_{G_k,0}](\boldsymbol{\alpha}_k)^* - [\Omega_k]^*[I_{G_k,0}](\boldsymbol{\omega}_k)^*, \quad (4.45)$$

$$(\mathbf{N}_{k-1})^* = [R(\theta_{k-1},\mathbf{u}_{k-1,0})]\{-(\mathbf{W}_k)^* - (\mathbf{F}_k)^* + (\mathbf{N}_k)^*\}, \quad (4.46)$$

$$\begin{aligned}(\mathbf{T}_{k-1})^* = [R(\theta_{k-1},\mathbf{u}_{k-1,0})]\{&-(\mathbf{c}_k)^*\times((\mathbf{W}_k)^*+(\mathbf{F}_k)^*) - (\mathbf{M}_k)^* \\ &+((\mathbf{c}_k)^*+(\mathbf{d}_k)^*)\times(\mathbf{N}_k)^* + (\mathbf{T}_k)^*\}. \quad (4.47)\end{aligned}$$

For the revolute jointed manipulator,

$$\begin{aligned}(\mathbf{c}_k)^* &= (\mathbf{P}^k_{k-1}\mathbf{G}_k)^* = \mathbf{c}_{k0}, \\ (\mathbf{d}_k)^* &= (\mathbf{G}_k\mathbf{P}^k_k)^* = \mathbf{d}_{k0},\end{aligned}$$

where $\mathbf{c}_{k0}$ and $\mathbf{d}_{k0}$ are the body vectors, in the ZRP position, from the $(k-1)$th joint center to the link mass center G_k, and from G_k

to the kth joint center, respectively. The situation is slightly more complicated for prismatic joints.

If the $(k-1)$th joint is prismatic, and the collar part belongs to link k, then we still have

$$(\mathbf{c}_k)^* = (\mathbf{P}_{k-1}^k \mathbf{G}_k)^* = \mathbf{c}_{k0},$$

but if the rod element belongs to link k, at the $(k-1)$th prismatic joint, then

$$(\mathbf{c}_k)^* = (\mathbf{P}_{k-1}^k \mathbf{G}_k)^* = \mathbf{c}_{k0} - s_{k-1}\mathbf{u}_{k-1,0}.$$

If the kth joint is prismatic, and the collar part belongs to link k, then we again have

$$(\mathbf{d}_k)^* = (\mathbf{G}_k \mathbf{P}_k^k)^* = \mathbf{d}_{k0},$$

but if the rod element belongs to link k, at the kth prismatic joint, then

$$(\mathbf{d}_k)^* = (\mathbf{G}_k \mathbf{P}_k^k)^* = \mathbf{d}_{k0} + s_k \mathbf{u}_{k0}.$$

The driving joint force f (for the prismatic joint) or torque τ (for the revolute joint) at the $(k-1)$th joint is found as

$$f_{k-1} = \mathbf{N}_{k-1} \cdot \mathbf{u}_{k-1} = (\mathbf{N}_{k-1})^* \cdot \mathbf{u}_{k-1,0} \quad \text{(prismatic joint)}, \tag{4.48}$$

$$\tau_{k-1} = \mathbf{T}_{k-1} \cdot \mathbf{u}_{k-1} = (\mathbf{T}_{k-1})^* \cdot \mathbf{u}_{k-1,0} \quad \text{(revolute joint)}, \tag{4.49}$$

where $(\mathbf{u}_{k-1})^* = \mathbf{u}_{k-1,0}$ has been used.

4.5 Lagrangian Formulation

In Lagrangian formulation, the internal forces at smooth connections or constraints do not appear. The accelerations are not needed, and the system can be considered as a whole (rather than considering individual free-body diagrams). Energy losses will be ignored for simplicity. The Lagrangian function is defined as $L = T - V$, where T is the kinetic energy of the system, and V is the potential energy of the system with respect to a convenient datum (e.g., the ZRP position). Previously, we have defined a 6×6 Jacobian matrix for the manipulator [Eq. (2.44)]. This can be considered as the Jacobian $[J_7]$ for the hand (body 7) motion. In a similar manner, we can define a $6 \times (i-1)$ sub-Jacobian matrix $[J_i]$ for the ith link that will relate link velocities $(\boldsymbol{\omega}_i, \mathbf{v}_{G_i})$ to $(i-1)$ joint rates; this link Jacobian will have $(i-1)$ columns of a form similar to those described for the hand Jacobian. The conversion to $(\boldsymbol{\omega}_i)^*$ and $(\mathbf{v}_{G_i})^*$ will be through the transpose of link rotation matrix, $[R_i^L]^t$. The kinetic energy is

$$
\begin{aligned}
T &= \sum_i (1/2) m_i (\mathbf{v}_{G_i})^t \mathbf{v}_{G_i} + \sum_i (1/2)(\boldsymbol{\omega}_i)^t [I_{G_i}] \boldsymbol{\omega}_i \\
&= \sum_i (1/2) m_i (\mathbf{v}_{G_i})^{*t} (\mathbf{v}_{G_i})^* + \sum_i (1/2)(\boldsymbol{\omega}_i)^{*t} [I_{G_i,0}] (\boldsymbol{\omega}_i)^* \\
&= (1/2) \sum_i \sum_j H_{ij} (d\theta_i/dt)(d\theta_j/dt),
\end{aligned}
\tag{4.50}
$$

where H_{ij} is a complicated function of joint θ's. The potential energy due to gravity is

$$
V = \sum_i m_i g (Z_{G_i} - Z_{G_i,0}),
\tag{4.51}
$$

where Z_{G_i} represents the Z coordinate of the mass center G_i of the ith link; the datum is chosen at the ZRP position. The Lagrange equation is

$$
\frac{d}{dt}\left(\frac{\partial L}{\partial \dot{q}_k}\right) - \frac{\partial L}{\partial q_k} = Q_k + \tau_k,
\tag{4.52}
$$

where τ_k is the kth torque and Q_k is the generalized force corresponding to all external forces $(\mathbf{F}_i^{\text{ext}})$ except gravity and all external moments $(\mathbf{M}_i^{\text{ext}})$ except joint torques.

$$
Q_k = \sum_{i=2}^{7} \mathbf{F}_i^{\text{ext}} \cdot \frac{\partial \mathbf{v}_i}{\partial \dot{q}_k} + \sum_{i=2}^{7} \mathbf{M}_i^{\text{ext}} \cdot \frac{\partial \boldsymbol{\omega}_i}{\partial \dot{q}_k}.
\tag{4.53}
$$

In terms of "$*$" quantities,

$$
Q_k = \sum_{i=2}^{7} \mathbf{F}_i^{*\text{ext}} \cdot \frac{\partial \mathbf{v}_i^*}{\partial \dot{q}_k} + \sum_{i=2}^{7} \mathbf{M}_i^{*\text{ext}} \cdot \frac{\partial \boldsymbol{\omega}_i^*}{\partial \dot{q}_k}.
\tag{4.54}
$$

The general form of the equations of motion, which describe system dynamics of the robot arm, is

$$
[H(\mathbf{Q})](d^2\mathbf{Q}/dt^2) + \mathbf{c}(\mathbf{Q}, d\mathbf{Q}/dt) + \mathbf{g}(\mathbf{Q}) = \mathbf{f},
\tag{4.55}
$$

where $\mathbf{Q}$ is a 6×1 vector of joint variables $\{q_i\}$—not to be confused with generalized forces Q_k above—$[H(\mathbf{Q})]$ is a 6×6 matrix, $\mathbf{c}(\mathbf{Q}, d\mathbf{Q}/dt)$ is a 6×1 vector containing centrifugal and Coriolis effects, $\mathbf{g}(\mathbf{Q})$ is a 6×1 vector containing gravity effects, and $\mathbf{f}$ is a 6×1 vector containing joint actuator forces (assuming that there are no other loading effects).

4.6 Control

Once the equations that describe the system dynamics of the robot arm become known, the robot can be controlled.

The components of a control system include error detectors, the controller unit, the "plant" or "system" to be controlled, potentiometers or optical shaft encoders (to measure position), tachometers (to measure velocity), A/D (analog-to-digital) and D/A (digital-to-analog) converters, sensors, and a microprocessor or a computer. A typical control system will have some or all of these components. A control system is described schematically by means of block diagrams; in fact, for classical linear control systems, a system of algebra, called block diagram algebra, has evolved from the work of Mason.

Let us discuss some common examples of automatic control. A thermostat-controlled home heating system is a familiar example. Once the thermostat is set, it measures the actual temperature and also serves as an error detector for the temperature. It provides a low-voltage on–off signal to the controller, which is an on–off electric relay. The relay turns on–off the high voltage blower–furnace unit. The hot air is distributed in the home through the duct system, and the temperature begins to rise and is sensed by the thermostat. This is an example of the on–off control system, and except for the initial sizing of the blower–furnace unit in relation to the size of the house, no information about the dynamic model of the "plant," which is blower–furnace–home, is utilized. The latest innovation in home heating is the use of a variable speed or multi-speed blower and more sophisticated electronic circuitry to actuate relays in order to provide a more gradual control action—high heat when the temperature difference is large, and low heat when it is small.

Other examples of control systems are automobile cruise control (similar in principle to the above example), airplane autopilot, servomechanisms, and numerically controlled (NC) machines. An unconventional example is the process of driving a car which involves man in the control loop. Eyes see the road ahead, detect the car's motion relative to the lanes of the road, and provide signals to the brain. The brain processes this information and commands the muscles of the hands to adjust the steering wheel, and the muscles of the legs to adjust the accelerator pedal (or to apply the brakes), to appropriately modify the gross movement or minor drifts of the car. This is an example of learning control, where the driver learns to drive the car through training and experience, and the knowledge of the inner workings of the human body or that of the car are not necessary for driving the car. This control system can also be called adaptive because once we learn how to drive, we can adapt easily to driving a variety of models with different weights, sizes, and operating characteristics.

A control system can have feedback loops as well as feedforward loops. The feedback loop is the most important feature of a control system. Through a feedback loop, the actual state of the system $(\mathbf{Q}_a, d\mathbf{Q}_a/dt)$ is measured and compared with the desired state $(\mathbf{Q}_d, d\mathbf{Q}_d/dt)$ of the system, and corrective action is taken by the controller based upon the position and velocity errors $(\varepsilon, d\varepsilon/dt)$, where $\varepsilon = \mathbf{Q}_d - \mathbf{Q}_a$. In this sense, feedback control is reactive. In the context of control, accelerations cannot be measured in a meaningful way—although there exist accelerometers that can measure the intensities of machine vibrations and earthquakes. It is also not practical to differentiate the velocity measurements because of the sensitivity of the differentiation process to errors and noises.

For simple dynamic systems, detailed information about the dynamic model is not needed to design practical feedback control systems. The feedback action based upon $(\varepsilon, d\varepsilon/dt)$ is obviously error driven, i.e., the corrective action is in reaction to the presence and detection of errors. A good controller unit will keep these errors within acceptable levels. However, in many complex systems, with nonlinear dynamics and other nonlinearities, the position and velocity feedback are not enough to achieve satisfactory control system performance.

A feedforward loop provides for some anticipatory action. Partial information about the desired state ($\mathbf{Q}_d$ or $d\mathbf{Q}_d/dt$ or $d^2\mathbf{Q}_d/dt^2$) is provided directly to the controller so that it can make anticipatory adjustments. In this sense, feedforward control "looks ahead." The feedforward action is supplemental to the feedback action, which is reactive. However, the degree of success with the feedforward loop depends upon the accuracy and reliability of the dynamic system model which is used by the controller to estimate "system" inputs. Determination of accurate system dynamics, and the calculations required to provide real-time estimation of system inputs, necessitate a computer or microprocessor in the loop, and that makes feedforward action an expensive and nontrivial proposition.

Another variation is to provide the actual state information ($\mathbf{Q}_a$, $d\mathbf{Q}_a/dt$), in addition to the usual error information ($\varepsilon, d\varepsilon/dt$), to the controller. The dynamic model of the system is utilized by the controller to estimate "system" inputs to maintain (or sustain) the actual position ($\mathbf{Q}_a$) and actual velocity ($d\mathbf{Q}_a/dt$) at a constant level (i.e., $d^2\mathbf{Q}_d/dt^2 = 0$). This is called nonlinear feedback. An intuitive way to think about this is that the "system" then acquires some desirable "inertial" characteristics, i.e., the "system" has a tendency to maintain a position of rest or a state of constant velocity. However, the main advantage of nonlinear feedback is that it can transform a nonlinear control system design problem into a linear control system design problem. Like the feedforward action, nonlinear feedback is also an expensive and nontrivial proposition; both require a computer in the loop to carry out real-time dynamics calculations.

A useful strategy for robot control involves a combination of position and velocity feedback, acceleration feedforward, and nonlinear feedback. The control law is developed as follows:

1. Position and velocity feedback are modified by $[H(\mathbf{Q}_a)]$.

$$\mathbf{f}' = [H(\mathbf{Q}_a)]([c_v](d\varepsilon/dt) + [c_p]\varepsilon). \tag{4.56}$$

   $[c_v]$ is a 6×6 velocity gain matrix and $[c_p]$ is a 6×6 position gain matrix.

2. Acceleration feedforward is modified by $[H(\mathbf{Q}_a)]$.

$$\mathbf{f}'' = [H(\mathbf{Q}_a)](d^2\mathbf{Q}_d/dt^2). \tag{4.57}$$

3. Nonlinear feedback is used to sustain $\mathbf{Q}_a$ and $(d\mathbf{Q}_a/dt)$ with $(d^2\mathbf{Q}_d/dt^2) = 0$.

$$\mathbf{f}''' = \mathbf{c}(\mathbf{Q}_a, d\mathbf{Q}_a/dt) + \mathbf{g}(\mathbf{Q}_a). \tag{4.58}$$

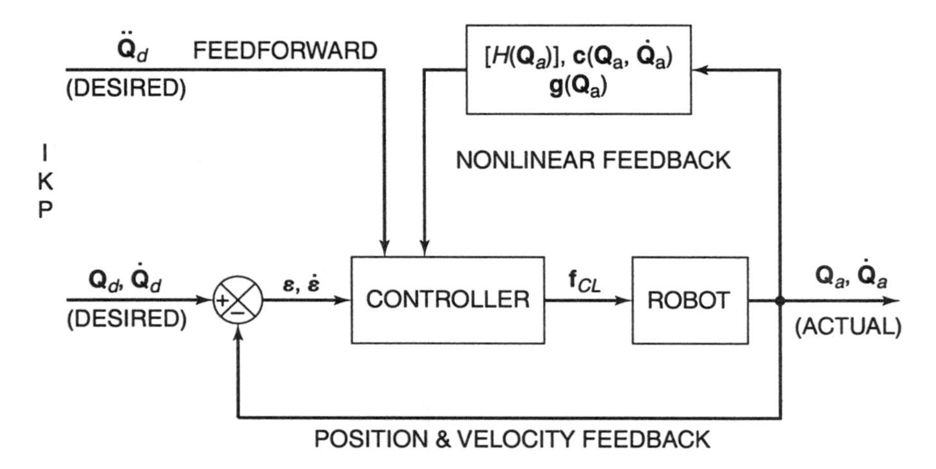

FIGURE 4.3. Control strategy for the computed torque method with feedforward and feedback (linear and nonlinear) loops indicated.

The control law, which defines the output of the controller, is

$$\mathbf{f}_{\mathrm{CL}} = \mathbf{f}' + \mathbf{f}'' + \mathbf{f}''' = [H(\mathbf{Q}_a)]((d^2\mathbf{Q}_d/dt^2) + [c_v](d\boldsymbol{\varepsilon}/dt) + [c_p]\boldsymbol{\varepsilon})$$
$$+ \mathbf{c}(\mathbf{Q}_a, d\mathbf{Q}_a/dt) + \mathbf{g}(\mathbf{Q}_a). \quad (4.59)$$

In implementation, real-time computations are necessary for $[H(\mathbf{Q}_a)]$, $\mathbf{c}(\mathbf{Q}_a, d\mathbf{Q}_a/dt)$, and $\mathbf{g}(\mathbf{Q}_a)$; the expressions are provided by the dynamic model, and the actual state $(\mathbf{Q}_a, d\mathbf{Q}_a/dt)$ is used for evaluation. The overall control strategy is shown in Figure 4.3.

Substituting the control force into system dynamics equations (4.55),

$$[H(\mathbf{Q}_a)](d^2\mathbf{Q}_a/dt^2) + \mathbf{c}(\mathbf{Q}_a, d\mathbf{Q}_a/dt) + \mathbf{g}(\mathbf{Q}_a) = \mathbf{f}_{\mathrm{CL}}. \quad (4.60)$$

Note that on the left-hand side, actual $\mathbf{Q}_a$, $(d\mathbf{Q}_a/dt)$, and $(d^2\mathbf{Q}_a/dt^2)$ are used in the system dynamics equations. The driving forces or torques at joints are provided through the control law [Eq. (4.59)] on the right-hand side. In hindsight, we can now see why the control force $\mathbf{f}_{\mathrm{CL}}$ was constructed as in Eq. (4.59). After substitution, nonlinear terms $\mathbf{c}(\mathbf{Q}_a, d\mathbf{Q}_a/dt)$ and $\mathbf{g}(\mathbf{Q}_a)$ cancel, and $[H(\mathbf{Q}_a)]$ can be factored out from the remainder to obtain

$$[H(\mathbf{Q}_a)]((d^2\boldsymbol{\varepsilon}/dt^2) + [c_v](d\boldsymbol{\varepsilon}/dt) + [c_p]\boldsymbol{\varepsilon}) = 0. \quad (4.61)$$

The error dynamics is then represented by six coupled, linear, second-order ordinary differential equations (ODEs):

$$(d^2\varepsilon/dt^2) + [c_v](d\varepsilon/dt) + [c_p]\varepsilon = 0. \tag{4.62}$$

These error dynamics equations are valid at all dynamic states of the manipulator and at any payload. If such a formulation is not used, then the control strategy (or law) would have to change with time as the manipulator moves or has different payloads. If the gain matrices are chosen as diagonal matrices, with diagonal elements c_{vi} for $[c_v]$ and c_{pi} for $[c_p]$, then we get six decoupled ODEs,

$$(d^2\varepsilon_i/dt^2) + c_{vi}(d\varepsilon_i/dt) + c_{pi}\varepsilon_i = 0, \quad i = 1, 2, \ldots, 6. \tag{4.63}$$

This is the simplest possible behavior for the joint errors. This simplicity is, however, deceptive because of the need to incorporate a computer or microprocessor in the control loop. Intuitively, we have modified the system through complex control loop computations in real-time and have effectively "linearized" it without making any approximations. In robotics applications, underdamped vibratory response is undesirable because when the hand approaches another object, it can lead to impacts. To provide critically damped response, choose the feedback gains as follows:

$$c_{vi} = 2(c_{pi})^{1/2}. \tag{4.64}$$

The control scheme above is sometimes referred to as the computed torque control scheme. The linearized error dynamics can be achieved only when the dynamic model is exact. However, the dynamic model is not exact because of manufacturing tolerances in link dimensions, experimental errors in the measurement or estimation of masses and inertias, and difficulties in accounting for the effects of electric wires, hydraulic hoses, etc. Studies have shown that if dimensional errors are less than 0.1%, and mass and inertia errors are less than 5%, then the control scheme above still gives satisfactory results.

The interested reader should consult specialized books on control theory to learn more about this subject.

Example 4.1. For the $R \perp R \perp P$ (spherical) regional structure shown in Figure 4.4, find the general expressions for link 4 velocities: $\mathbf{v}_{G4}$, $\boldsymbol{\omega}_4$, and star quantity $(\boldsymbol{\omega}_4)^*$ as functions of the joint angles

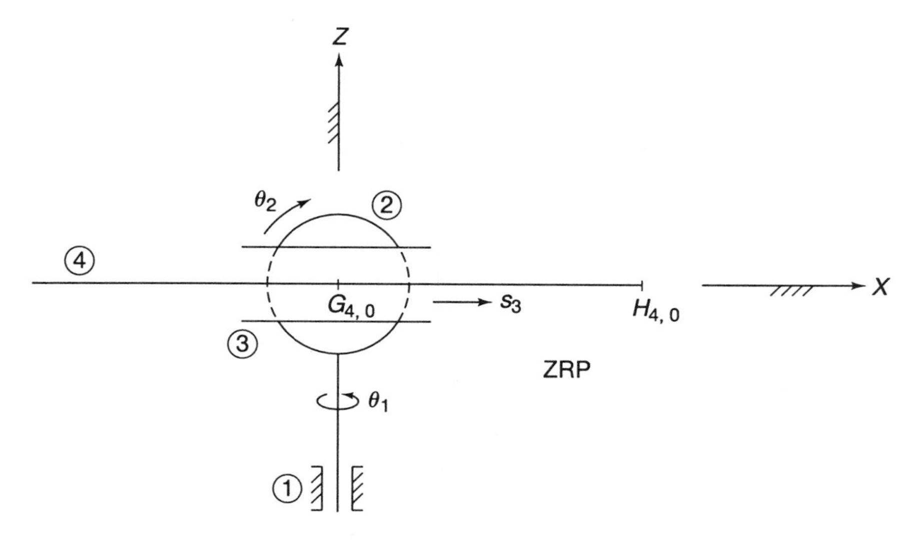

FIGURE 4.4. ZRP position for the spherical regional structure used in Example 4.1.

and rates. Ignore the masses and inertias of links 2 and 3 (base is link 1). Assume that link 4 has an approximately cylindrical shape, with the axial moment of inertia I_{a4} and the transverse moments of inertia I_{t4} with respect to its centroid G_4. Consider that only external loads are gravitational. Derive the general expressions for the joint torques/forces τ_1, τ_2, f_3 from the Lagrange equations. Finally, develop the appropriate control law for the computed torque scheme.

Solution: The main purpose of this example is to show the use of some of the equations that have been developed in the previous section. Without the use of simplifying assumptions for masses and inertias, computer programs will be required to perform numerical calculations or algebraic manipulations.

The link velocities can be obtained in two ways, and both of these will be outlined here. In the first approach, velocities $\boldsymbol{\omega}_4$ and $\mathbf{v}_{G4}$ are related to the joint rates $d\theta_1/dt$, $d\theta_2/dt$, and ds_3/dt by means of a 6×3 link sub-Jacobian $[J_4]$ with a form similar to that in Eq. (2.44) except that it has only three columns. Once $\boldsymbol{\omega}_4$ is determined in this manner, $(\boldsymbol{\omega}_4)^*$ is found from its definition (4.27): $(\boldsymbol{\omega}_4)^* = [R_4^L]^t\boldsymbol{\omega}_4$, where $[R_4^L]$ is the rotation matrix for link 4.

$$\begin{bmatrix} \boldsymbol{\omega}_4 \\ \mathbf{v}_{G4} \end{bmatrix} = \begin{bmatrix} \mathbf{u}_1 & \mathbf{u}_2 & \mathbf{0} \\ \mathbf{u}_1 \times \mathbf{Q}_1\mathbf{G}_4 & \mathbf{u}_2 \times \mathbf{Q}_2\mathbf{G}_4 & \mathbf{u}_3 \end{bmatrix}_{6\times 3} \begin{bmatrix} d\theta_1/dt \\ d\theta_2/dt \\ ds_3/dt \end{bmatrix}.$$

Note that vector $\mathbf{Q}_i$ represents the position vector of a point on the ith joint axis. The joint displacement matrices are

$$D_1 = D(\theta_1, 0, \mathbf{K}, \mathbf{0}), \quad D_2 = D(\theta_2, 0, \mathbf{J}, \mathbf{0}), \quad D_3 = D(0, s_3, \mathbf{I}, \mathbf{0}),$$

and the joint rotation matrices (3×3 principal minors of above) are

$$R_1 = R(\theta_1, \mathbf{K}), \quad R_2 = R(\theta_2, \mathbf{J}), \quad R_3 = R(0, \mathbf{I}) = I_{3\times 3}.$$

Note that

$$\mathbf{Q}_1 = \mathbf{Q}_2 = \mathbf{Q}_{1,0} = \mathbf{Q}_{2,0} = (0,0,0)^t,$$
$$\mathbf{u}_1 = \mathbf{u}_{1,0} = (0,0,1)^t, \quad \mathbf{u}_{2,0} = (0,1,0)^t, \quad \mathbf{u}_{3,0} = (1,0,0)^t.$$

We find the current directions for joint axes 2 and 3, i.e., $\mathbf{u}_2$ and $\mathbf{u}_3$, as

$$\mathbf{u}_2 = R_1\mathbf{u}_{2,0} = \begin{bmatrix} -\sin\theta_1 \\ \cos\theta_1 \\ 0 \end{bmatrix},$$

$$\mathbf{u}_3 = R_1R_2\mathbf{u}_{3,0} = \begin{bmatrix} \cos\theta_1\cos\theta_2 \\ \sin\theta_1\cos\theta_2 \\ -\sin\theta_2 \end{bmatrix}.$$

The link displacement matrix for link 4 is

$$\begin{aligned} D_4^L &= D_1D_2D_3 \\ &= \begin{bmatrix} \cos\theta_1\cos\theta_2 & -\sin\theta_1 & \cos\theta_1\sin\theta_2 & s_3\cos\theta_1\cos\theta_2 \\ \sin\theta_1\cos\theta_2 & \cos\theta_1 & \sin\theta_1\sin\theta_2 & s_3\sin\theta_1\cos\theta_2 \\ -\sin\theta_2 & 0 & \cos\theta_2 & -s_3\sin\theta_2 \\ 0 & 0 & 0 & 1 \end{bmatrix}, \end{aligned}$$

and the corresponding link rotation matrix is

$$R_4^L = R_1R_2R_3 = \begin{bmatrix} \cos\theta_1\cos\theta_2 & -\sin\theta_1 & \cos\theta_1\sin\theta_2 \\ \sin\theta_1\cos\theta_2 & \cos\theta_1 & \sin\theta_1\sin\theta_2 \\ -\sin\theta_2 & 0 & \cos\theta_2 \end{bmatrix}.$$

The current position of mass center $\mathbf{G}_4$ is obtained from $\mathbf{G}_{4,0} = (0,0,0)^t$ as

$$\begin{bmatrix} \mathbf{G}_4 \\ 1 \end{bmatrix} = D_4^L \begin{bmatrix} \mathbf{G}_{4,0} \\ 1 \end{bmatrix} = \begin{bmatrix} s_3 \cos\theta_1 \cos\theta_2 \\ s_3 \sin\theta_1 \cos\theta_2 \\ -s_3 \sin\theta_2 \\ 1 \end{bmatrix}.$$

Upon substitution in the expression for $[J_4]$, and expanding the cross products, we get

$$[J_4] = \begin{bmatrix} 0 & -\sin\theta_1 & 0 \\ 0 & \cos\theta_1 & 0 \\ 1 & 0 & 0 \\ -s_3 \sin\theta_1 \cos\theta_2 & -s_3 \cos\theta_1 \sin\theta_2 & \cos\theta_1 \cos\theta_2 \\ s_3 \cos\theta_1 \cos\theta_2 & -s_3 \sin\theta_1 \sin\theta_2 & \sin\theta_1 \cos\theta_2 \\ 0 & -s_3 \cos\theta_2 & -\sin\theta_2 \end{bmatrix}_{6\times 3}.$$

The velocities for link 4 can now be obtained.

$$\boldsymbol{\omega}_4 = \begin{bmatrix} -(d\theta_2/dt)\sin\theta_1 \\ (d\theta_2/dt)\cos\theta_1 \\ d\theta_1/dt \end{bmatrix},$$

$$\mathbf{v}_{G4} =$$

$$\begin{bmatrix} -(d\theta_1/dt)s_3 \sin\theta_1 \cos\theta_2 - (d\theta_2/dt)s_3 \cos\theta_1 \sin\theta_2 \\ + (ds_3/dt)\cos\theta_1 \cos\theta_2 \\ (d\theta_1/dt)s_3 \cos\theta_1 \cos\theta_2 - (d\theta_2/dt)s_3 \sin\theta_1 \sin\theta_2 \\ + (ds_3/dt)\sin\theta_1 \cos\theta_2 \\ -(d\theta_2/dt)s_3 \cos\theta_2 - (ds_3/dt)\sin\theta_2 \end{bmatrix}.$$

From the definition of $(\boldsymbol{\omega}_4)^*$,

$$(\boldsymbol{\omega}_4)^* = [R_4^L]^t \boldsymbol{\omega}_4 = \begin{bmatrix} -(d\theta_1/dt)\sin\theta_2 \\ d\theta_2/dt \\ (d\theta_1/dt)\cos\theta_2 \end{bmatrix}.$$

Although conceptually appealing, the above method for finding $(\boldsymbol{\omega}_4)^*$ is quite labor-intensive. The second approach is to use the recursion in Eq. (4.35),

$$(\boldsymbol{\omega}_{k+1})^* = [R(\theta_k, \mathbf{u}_{k0})]^t \{(\boldsymbol{\omega}_k)^* + (d\theta_k/dt)\mathbf{u}_{k0}\},$$

or, more simply,

$$(\boldsymbol{\omega}_{k+1})^* = [R(\theta_k, \mathbf{u}_{k0})]^t(\boldsymbol{\omega}_k)^* + (d\theta_k/dt)\mathbf{u}_{k0}. \tag{4.35}$$

Starting with $k = 1$ and $(\boldsymbol{\omega}_1)^* = \mathbf{0}$, and using $\mathbf{J} = (0,1,0)^t$ and $\mathbf{K} = (0,0,1)^t$,

$$\begin{aligned}(\boldsymbol{\omega}_2)^* &= [R(\theta_1, \mathbf{u}_{1,0})]^t(\boldsymbol{\omega}_1)^* + (d\theta_1/dt)\mathbf{u}_{1,0} = (d\theta_1/dt)\mathbf{K}\\ (\boldsymbol{\omega}_3)^* &= [R(\theta_2, \mathbf{u}_{2,0})]^t(\boldsymbol{\omega}_2)^* + (d\theta_2/dt)\mathbf{u}_{2,0}\\ &= (d\theta_1/dt)[R(\theta_2, \mathbf{J})]^t\mathbf{K} + (d\theta_2/dt)\mathbf{J}\end{aligned}$$

or

$$(\boldsymbol{\omega}_3)^* = \begin{bmatrix} -(d\theta_1/dt)\sin\theta_2 \\ d\theta_2/dt \\ (d\theta_1/dt)\cos\theta_2 \end{bmatrix}.$$

Finally, since the third joint is prismatic [see Eq. (4.41)],

$$(\boldsymbol{\omega}_4)^* = (\boldsymbol{\omega}_3)^* = \begin{bmatrix} -(d\theta_1/dt)\sin\theta_2 \\ d\theta_2/dt \\ (d\theta_1/dt)\cos\theta_2 \end{bmatrix},$$

which is the same result. Although the expressions have not been given here for the recursion of $(\mathbf{v}_{G_i})^*$, it can also be obtained quickly in this way (see Problem 5). However, since $\mathbf{v}_{G_4}$ is already available, in this solution, we will use it in the subsequent derivations. The kinetic energy T is expressed as

$$\begin{aligned}T &= (1/2)m_4(\mathbf{v}_{G_4})^t\mathbf{v}_{G_4} + (1/2)(\boldsymbol{\omega}_4)^t[I_{G_4}]\boldsymbol{\omega}_4\\ &= (1/2)m_4(\mathbf{v}_{G_4})^t(\mathbf{v}_{G_4}) + (1/2)(\boldsymbol{\omega}_4)^{*t}[I_{G_4,0}](\boldsymbol{\omega}_4)^*,\end{aligned}$$

where the constant inertia matrix is

$$[I_{G_4,0}] = \begin{bmatrix} I_{a4} & 0 & 0 \\ 0 & I_{t4} & 0 \\ 0 & 0 & I_{t4} \end{bmatrix}.$$

After simplification, the kinetic energy expression becomes

$$\begin{aligned}T &= (1/2)m_4(ds_3/dt)^2 + (1/2)I_{a4}(d\theta_1/dt)^2(\sin\theta_2)^2\\ &\quad + (1/2)(I_{t4} + m_4 s_3^2)[(d\theta_1/dt)^2(\cos\theta_2)^2 + (d\theta_2/dt)^2].\end{aligned}$$

The potential energy expression, with datum at the ZRP position, becomes

$$V = m_4 g(Z_{G_4} - Z_{G_4,0}) = m_4 g Z_{G_4} = -m_4 g s_3 \sin\theta_2.$$

The Lagrangian is $L = T - V$. Using the Lagrange equations,

$$\frac{d}{dt}\left(\frac{\partial L}{\partial \dot{q}_k}\right) - \frac{\partial L}{\partial q_k} = Q_k + \tau_k.$$

To illustrate, for $k = 3$,

$$\begin{aligned}
\frac{\partial T}{\partial \dot{s}_3} &= m_4 \dot{s}_3, \quad \frac{d}{dt}\left(\frac{\partial T}{\partial \dot{s}_3}\right) = m_4 \ddot{s}_3, \\
\frac{\partial T}{\partial s_3} &= m_4 s_3 \{\dot{\theta}_1^2 (\cos\theta_2)^2 + \dot{\theta}_2^2\}, \quad \frac{\partial V}{\partial s_3} = -m_4 g \sin\theta_2, \\
f_3 &= m_4 \ddot{s}_3 - m_4 s_3 \{\dot{\theta}_1^2 (\cos\theta_2)^2 + \dot{\theta}_2^2\} - m_4 g \sin\theta_2.
\end{aligned}$$

After much simplification, we obtain

$$[H(\mathbf{Q})](d^2\mathbf{Q}/dt^2) + \mathbf{c}(\mathbf{Q}, d\mathbf{Q}/dt) + \mathbf{g}(\mathbf{Q}) = \mathbf{f},$$

where $\mathbf{Q} = (\theta_1, \theta_2, s_3)^t$, not to be confused with the position vector $\mathbf{Q}_i$ above or with generalized forces Q_k ($= 0$ here), and actuation "force" $\mathbf{f} = (\tau_1, \tau_2, f_3)^t$. The elements of the 3×3 matrix $[H(\mathbf{Q})]$, diagonal in this case, are

$$\begin{aligned}
H_{11} &= I_{a4} (\sin\theta_2)^2 + (I_{t4} + m_4 s_3^2)(\cos\theta_2)^2, \\
H_{22} &= I_{t4} + m_4 s_3^2, \\
H_{33} &= m_4.
\end{aligned}$$

The elements of the 3×1 vector $\mathbf{c}(\mathbf{Q}, d\mathbf{Q}/dt)$ are

$$\begin{aligned}
c_1 &= (I_{a4} - I_{t4} - m_4 s_3^2)(d\theta_1/dt)(d\theta_2/dt) \sin 2\theta_2 \\
&\quad + 2m_4 (d\theta_1/dt)(ds_3/dt) s_3 (\cos\theta_2)^2, \\
c_2 &= -(I_{a4} - I_{t4} - m_4 s_3^2)(d\theta_1/dt)^2 \sin\theta_1 \cos\theta_1 \\
&\quad + 2m_4 s_3 (d\theta_2/dt)(ds_3/dt), \\
c_3 &= -m_4 (d\theta_1/dt)^2 s_3 (\cos\theta_2)^2 - m_4 s_3 (d\theta_2/dt)^2.
\end{aligned}$$

The elements of the 3×1 vector $\mathbf{g}(\mathbf{Q})$ are

$$\begin{aligned}
g_1 &= 0, \\
g_2 &= -m_4 g s_3 \cos\theta_2, \\
g_3 &= -m_4 g \sin\theta_2.
\end{aligned}$$

The error is defined as

$$\boldsymbol{\varepsilon} = \mathbf{Q}_d - \mathbf{Q}_a = \begin{bmatrix} \theta_{1,d} - \theta_{1,a} \\ \theta_{2,d} - \theta_{2,a} \\ s_{3,d} - s_{3,a} \end{bmatrix}.$$

The control law for the computed torque scheme is

$$\mathbf{f}_{\text{CL}} = [H(\mathbf{Q}_a)]((d^2\mathbf{Q}_d/dt^2) + [c_v](d\boldsymbol{\varepsilon}/dt) + [c_p]\boldsymbol{\varepsilon})$$
$$+ \mathbf{c}(\mathbf{Q}_a, d\mathbf{Q}_a/dt) + \mathbf{g}(\mathbf{Q}_a).$$

Even for this simple case, which corresponds to a heavy body 4 with 3 dof (2R, 1P), the equations of motion and the control law become complicated. Such closed-form derivations are not manually feasible for general 6 dof robots, and computer methods must be used to perform these computations numerically or algebraically.

4.7 Problems

1. Derive Eq. (4.23) for the inertia moments in the principal centroidal body system.

2. From the nonbody forms of the equations of motion, Eqs. (4.6) and (4.26), derive the ZRP forms of the Newton–Euler equations in (4.28) and (4.29).

3. Derive the forward and backward recursions for modified angular velocity $(\boldsymbol{\omega})^*$ in Eqs. (4.35) and (4.37).

4. Derive the forward recursion formula for the modified angular acceleration $(\boldsymbol{\alpha})^*$ in Eq. (4.36).

5. Derive the following recursion relations for modified linear velocities:

$$(\mathbf{v}_{P_k^k})^* = (\mathbf{v}_{G_k})^* + (\boldsymbol{\omega}_k)^* \times (\mathbf{d}_k)^*$$
$$(\mathbf{v}_{P_k^{k+1}})^* = [R(\theta_k, \mathbf{u}_{k0})]^t (\mathbf{v}_{P_k^k})^* \quad \text{(joint } k \text{ is revolute)}$$
$$(\mathbf{v}_{P_k^{k+1}})^* = (\mathbf{v}_{P_k^k})^* + (ds_k/dt)\mathbf{u}_{ko} \quad \text{(joint } k \text{ is prismatic)}$$
$$(\mathbf{v}_{G_{k+1}})^* = (\mathbf{v}_{P_k^{k+1}})^* + (\omega_{k+1})^* \times (\mathbf{c}_{k+1})^*.$$

6. Derive the following recursion relations for modified acceleration:

$$(\mathbf{a}_{P_k^k})^* = (\mathbf{a}_{G_k})^* + (\boldsymbol{\omega}_k)^* \times \{(\boldsymbol{\omega}_k)^* \times (\mathbf{d}_k)^*\} + (\boldsymbol{\alpha}_k)^* \times (\mathbf{d}_k)^*$$

$$(\mathbf{a}_{P_k^{k+1}})^* = [R(\theta_k, \mathbf{u}_{k0})]^t (\mathbf{a}_{P_k^k})^* \quad \text{(joint } k \text{ is revolute)}$$

$$(\mathbf{a}_{P_k^{k+1}})^* = (\mathbf{a}_{P_k^k})^* + (d^2 s_k/dt^2)\mathbf{u}_{ko} + 2(ds_k/dt)(\boldsymbol{\omega}_k)^* \times \mathbf{u}_{ko}$$

$$\text{(joint } k \text{ is prismatic)}$$

$$\begin{aligned}(\mathbf{a}_{G_{k+1}})^* &= (\mathbf{a}_{P_k^{k+1}})^* + (\omega_{k+1})^* \times \{(\omega_{k+1})^* \times (\mathbf{c}_{k+1})^*\} \\ &\quad + (\alpha_{k+1})^* \times (\mathbf{c}_{k+1})^*.\end{aligned}$$

7. A 2-dof manipulator is shown at its zero-reference-position in Figure P4.1.

   (a) Use the star-superscripted quantities to find the general expressions for modified velocities $(\boldsymbol{\omega}_3)^*$ and $(\mathbf{v}_{G_3})^*$. Assume that link 3 is much heavier than link 2 and thus ignore the mass and inertia of link 2. For link 3, use mass m_3 and the inertia matrix at ZRP as

$$I_{G3,0} = \begin{bmatrix} I_x & 0 & 0 \\ 0 & I_y & 0 \\ 0 & 0 & I_z \end{bmatrix}.$$

   (b) Use the Lagrange equation to find the general expressions for joint torques τ_1 and τ_2.

   (c) Write the control law that will render the joint error dynamic equations as linear decoupled ODEs.

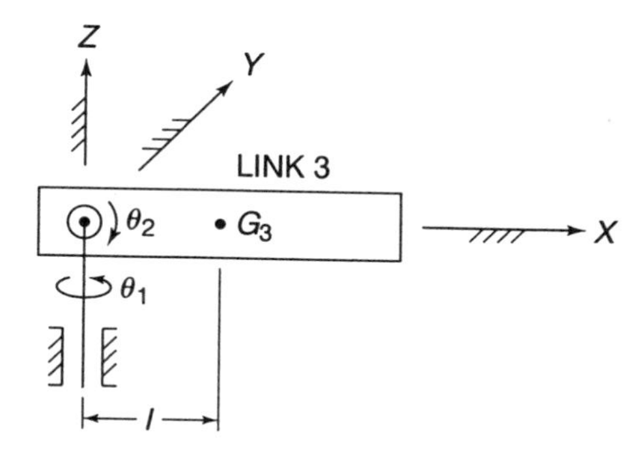

FIGURE P4.1. ZRP position for the 2-dof manipulator in Problem 7.

8. A one-armed robot has the following equation of motion:

$$I_o(d^2\theta/dt^2) + 0.1\, mL^2(d\theta/dt)^2 + 0.5\, mgL\cos\theta = \tau.$$

Formulate the expression for control torque τ_{CL} for the "computed torque scheme." If the gain c_p is chosen to be 225.0 sec^{-2}, then what should be the value of the gain c_v?

9. A 2 dof robot has the following dynamical equations:

$$(10\cos^2\theta_2)(d^2\theta_1/dt^2) + (10\sin 2\theta_2)(d\theta_1/dt)(d\theta_2/dt) = \tau_1 \quad (\text{ft}-\text{lb})$$

$$10(d^2\theta_2/dt^2) + (5\sin 2\theta_2)(d\theta_1/dt)^2 - 50\cos\theta_2 = \tau_2. \quad (\text{ft}-\text{lb})$$

The desired joint values and rates, as determined from the inverse kinematics processor, are

$$\theta_d = (29^\circ, 46^\circ)^t, \quad (d\theta_d/dt) = (2.0, 5.5)^t\text{rad/s},$$

$$(d^2\theta_d/dt^2) = (10.0, 15.0)^t\text{rad/s}^2.$$

The actual (measured) joint values and velocities are

$$\theta_a = (30^\circ, 45^\circ)^t, \quad (d\theta_a/dt) = (2.1, 5.7)^t\text{rad/s}.$$

Based upon the control law that includes position and velocity feedback (use $c_{pi} = 100\ \text{sec}^{-2}$, $c_{vi} = 20\ \text{sec}^{-1}$, $i = 1, 2$, and change θ from degrees to radians), acceleration feedforward and nonlinear dynamical feedback, calculate the control torque τ_{control} that should be generated through the controller commands.

References

Angeles, J., On the numerical solution for the inverse kinematics problem, *International Journal of Robotics Research*, **4**(2):21–37, 1985.

Armstrong, B., Khatib, O., and Burdick, J., Explicit dynamic model and inertial parameters of the PUMA 560 arm, *Proceedings of the IEEE Conference on Robotics and Automation*, San Francisco, 510–518, 1986.

Asada, H. and Granito, J.A.C., Kinematic and static characterization of wrist joints and their optimal design, *Proceedings of the IEEE Conference on Robotics and Automation*, St. Louis, 244–250, 1985.

Asada, H. and Slotine, J., *Robot Analysis and Control*, Wiley, New York, 1986.

Chen, C.L., Lee, C.S.G., and Hou, E.S.H., Efficient scheduling algorithm for robot inverse dynamics computation on a multiprocessor system, *Proceedings of the IEEE Conference on Robotics and Automation*, Philadelphia, **2**:1146–1151, 1988.

Chen, X. and Gupta, K.C., Geometric modeling and visualization of manipulator workspace, *Proceedings of the ASME Computers in Engineering Conference*, **1**:469–474, 1991.

Cheng, H. and Gupta, K.C., A study of robot inverse kinematics based upon the solution of differential equations, *Journal of Robotic Systems*, **8**(2):159–175, 1991.

Cheng, H.H. and Gupta, K.C., An efficient manipulator dynamics based upon Newton–Euler equations and the ZRP method, *Proceedings of the ASME Mechanisms Conference*, Scottsdale, **47**:81–87, 1992.

Cheng, H.H. and Gupta, K.C., Vectorization of robot inverse dynamics on a pipelined vector processor, *IEEE Transactions on Robotics and Automation*, **9**(6):858–863, 1993.

Craig, J.J., *Introduction to Robotics—Mechanics and Control*, Addison–Wesley, Reading, MA, 1986.

Craig, J.J., *Adaptive Control of Mechanical Manipulators*, Addison–Wesley, Reading, MA, 1988.

Crandall, S.H., Karnopp, D.C., Kurtz, E.F., and Pridmore-Brown, C., *Dynamics of Mechanical and Electromechanical Systems*, McGraw–Hill, New York, 1968.

Davidson, J.K. and Hunt, K.H., Rigid body location and robot workspaces: Some alternative manipulator forms, *ASME Journal of Mechanisms, Transmissions, and Automation in Design*, **109**:224–232, 1987.

Denavit, J. and Hartenberg, R.S., A kinematic notation for lower pair mechanisms based on matrices, *ASME Journal of Applied Mechanics*, **6**:215–221, 1955.

Desa, S. and Roth, B., Mechanics—Kinematics and Dynamics, *Recent Advances in Robotics*, 71–130, (eds.) Beni, G. and Hackwood, S., Wiley, New York, 1985.

Duffy, J., *Analysis of Mechanisms and Manipulators*, Wiley, New York, 1980.

Duffy, J. and Crane, C., A displacement analysis of the general spatial 7R mechanism, *Mechanism and Machine Theory*, **15**:153–169, 1980.

Featherstone, R., Position and velocity transformations between end-effector coordinates and joint angles, *International Journal of Robotics Research*, **2**:35–45, 1983.

Fichter, E.F. and Hunt, K.H., The fecund torus—its bitangent-circles and derived linkages, *Mechanism and Machine Theory*, **10**:167–176, 1975.

Freudenstein, F., Longman, R.W., and Chen, C.K., Kinematic analysis of robotic bevel gear trains, *ASME Journal of Mechanisms, Transmissions, and Automation in Design*, **106**:371–375, 1984.

Freudenstein, F. and Primrose, E.J.F., On the analysis and synthesis of the workspace of a three-link, turing-pair connected robot arm, *ASME Journal of Mechanisms, Transmissions, and Automation in Design*, **106**:365–370, 1984.

Gilbert, E.G. and Ha, I.J., An approach to nonlinear feedback control with applications to robotics, *IEEE Conference on Robotics and Automation*, Atlanta, 879–884, 1984.

Goldenberg, A.A. and Lawrence, D.L., A generalized solution to the inverse kinematics of robot manipulators, *ASME Journal of Dynamic Systems, Measurements, and Control*, **107**:103–106, 1985.

Groover, M.P., Weiss, M., Nagel, R.N., and Odrey, N.G., *Industrial Robotics—Technology, Programming, and Applications*, McGraw–Hill, New York, 1986.

Gupta, K.C., A note on position analysis of manipulators, *Proceedings of the Applied Mechanisms Conference*, Kansas City, II.1–3, 1981; see also, *Mechanism and Machine Theory*, **19**:5–8, 1984.

Gupta, K.C., Discussion on kinematic analysis of robotic bevel gear trains, *ASME Journal of Mechanisms, Transmissions, and Automation in Design*, **107**:142–143, 1985.

Gupta, K.C., Discussion on a study of the Jacobian matrix of serial manipulators, *ASME Journal of Mechanisms, Transmissions, and Automation in Design*, **107**:237–238, 1985.

Gupta, K.C., Kinematic analysis of manipulators using the zero reference position description, *International Journal of Robotics Research*, **5**(2):5–13, 1986.

Gupta, K.C., On the nature of workspace, *International Journal of Robotics Research*, **5**(2):112–121, 1986.

Gupta, K.C., Rotatability considerations for spherical four-bar linkages with applications to robot wrist design, *ASME Journal of Mechanisms, Transmissions, and Automation in Design*, **108**:387–391, 1986.

Gupta, K.C., Kinematics of a robot with continuous roll wrist, *IEEE Transactions on Robotics and Automation*, **4**(4):440–443, 1988.

Gupta, K. C. and Carlson, G.J., On certain aspects of the zero reference position method and its application to an industrial manipulator, *Journal of Robotic Systems*, **3**(1):41–57, 1986.

Gupta, K.C. and Kazerounian, S.M.K., Improved numerical solutions of inverse kinematics of robots, *Proceedings of the IEEE Conference on Robotics and Automation*, St. Louis, 743–748, 1985.

Gupta, K.C. and Ma, R., Formulation of manipulator Jacobian matrices using the velocity similarity principle, *Robotica*, **8**:81–84, 1990.

Gupta, K.C. and Ma, R., A direct rotatability criterion for spherical four-bar linkages, *ASME Journal of Mechanical Design*, **117**:597–600, 1995.

Gupta, K.C. and Ma, R., Kinematics of robot wrists and related linkages, *Proceedings of the IFToMM World Congress on the Theory of Machines and Mechanisms*, Italy, **3**:1930–1934, 1995.

Gupta, K.C. and Roth, B., Design considerations for manipulator workspace, *ASME Journal of Mechanical Design*, **104**:704–711, 1982.

Gupta, K.C. and Singh, V.K., A numerical algorithm for solving robot inverse kinematics, *Robotica*, **7**:159–164, 1989.

Hansen, J.A., Gupta, K.C., and Kazerounian, S.M.K., Generation and evaluation of the workspace of a manipulator, *International Journal of Robotics Research*, **2**:22–31, 1983.

Hartenberg, R.S. and Denavit, J., *Kinematic Synthesis of Linkages*, McGraw–Hill, New York, 1964.

Hashimoto, K. and Kimura, H., A new parallel algorithm for inverse dynamics, *International Journal of Robotics Research*, **8**(1):63–76, 1989.

Herve, J.M., Intrinsic formulation of problems of geometry and kinematics of mechanisms, *Mechanism and Machine Theory*, **17**:179–184, 1982.

Hollerbach, J.M., A recursive Lagrangian formulation of manipulator dynamics and a comparative study of dynamics formulation complexity, *IEEE Transactions on Systems, Man, and Cybernetics*, **10**(11):730–736, 1980.

Hollerbach, J.M. and Sahar, G., Wrist-partitioned, inverse kinematic accelerations and manipulator dynamics, *International Journal of Robotics Research*, **2**(4):61–76, 1983.

Hunt, K.H., Robot kinematics—A compact analytic inverse solution for velocities, *ASME Journal of Mechanisms, Transmissions, and Automation in Design*, **109**(1):42–49, 1987.

Huston, R.L. and Kelly, F.A., Development of equations of motion of single-arm robots, *IEEE Transactions on Systems, Man, and Cybernetics*, **12**(3):259–266, 1982.

Ichiro, K., *Mechanical Hands Illustrated*, Hemisphere Publishing, New York, 1987.

Kane, T.R. and Levinson, D.A., The use of Kane's dynamical equations in robotics, *International Journal of Robotics Research*, **2**(3):3–20, 1983.

Kazerounian, K. and Gupta, K.C., Manipulator dynamics using the extended zero position description, *Proceedings of the Applied Mechanisms Conference*, Kansas City, **2**:6.1–6.11, 1985; see also *IEEE Transactions on Robotics and Automation*, **2**(4):221–224, 1986.

Kazerounian, S.M.K., On the numerical inverse kinematics of robot manipulators, *ASME Journal of Mechanisms, Transmissions, and Automation in Design*, **109**:8–13, 1987.

Khosla, P.K. and Kanade, T., Real-time implementation and evaluation of model-based control on CMU DD (Direct Drive) arm II, *IEEE Conference on Robotics and Automation*, San Francisco, 1986.

Khosla, P.K. and Kanade, T., Experimental evaluation of nonlinear feedback and feedforward control schemes for manipulators, *International Journal of Robotics Research*, **7**(1):18–28, 1988.

Klafter, R.D., Chmielewski, T.A., and Negin, M., *Robotic Engineering—An Integrated Approach*, Prentice–Hall, Englewood Cliffs, NJ, 1989.

Kohli, D. and Hsu, M.S., The Jacobian analysis of workspace of mechanical manipulators, *Mechanism and Machine Theory*, **22**:265–275, 1987; see also pages 277–289 for a related paper by the same authors.

Kohli, D. and Osvatic, M., Inverse kinematics of general 6R and 5R-P serial manipulators, *ASME Journal of Mechanical Design*, **115**(4):922–931, 1993.

Kohli, D. and Spanos, J., Workspace analysis of mechanical manipulators using polynomial discriminants, *ASME Journal of Mecha-*

nisms, Transmissions, and Automation in Design, **107**:209–215, 1985; see also pages 216–222 for a related work by the same authors.

Kumar, A. and Patel, M.S., Mapping the manipulator workspace using interactive computer graphics, *International Journal of Robotics Research*, **5**:122–130, 1986.

Kumar, A. and Waldron, K.J., The workspace of a mechanical manipulator, *ASME Journal of Mechanical Design*, **103**:665–672, 1981.

Lathrop, R.H., Parallelism in manipulator dynamics, *IEEE Conference on Robotics and Automation*, St. Louis, 772–778, 1985.

Lee, C.S.G., Robot arm kinematics, dynamics and control, *Computer*, **15**:62–80, 1982.

Lee, C.S.G., Gonzalez, R.C., and Fu, K.S., *Tutorial on Robotics*, IEEE Computer Society, New York, 1983.

Lee, T.W. and Yang, D.C.H., On the evaluation of manipulator workspace, *ASME Journal of Mechanisms, Transmissions, and Automation in Design*, **105**:70–77, 1983.

Lenarcic, J., An efficient numerical approach for calculating the inverse kinematics for robot manipulators, *Robotica*, **3**:21–26, 1985.

Leu, M.C. and Hemati, N., Automated symbolic derivation of dynamic equations of motion for robot manipulators, *ASME Journal of Dynamic Systems, Measurements, and Control*, **108**:172–179, 1986.

Li, H.Y. and Liang, C.G., Displacement analysis of general spatial 7-link 7R mechanism, *Mechanism and Machine Theory*, **23**(3):219–226, 1988.

Li, H.Y., Woernle, C., and Hiller, M., A complete solution for the inverse kinematic problem of the general 6R robot manipulator, *Proceedings of the 1990 ASME Mechanisms Conference*, Chicago, DE**25**:45–52, 1990.

Lin, C.C.D. and Freudenstein, F., Optimization of the workspace of a three-link turning-pair connected robot arm, *International Journal of Robotics Research*, **5**:104–111, 1986.

Lin, P.N. and Duffy, J., The mapping and structure of the workspace of robot manipulators with revolute and prismatic pairs, *Proceedings of the ASME Computers in Engineering Conference*, San Diego, **2**:165–172, 1982.

Litvin, F.L. and Parenti-Castelli, V., Robot's manipulators: Simulation and identification of configurations, execution of prescribed trajectories, *Proceedings of the IEEE Conference on Robotics and Automation*, Atlanta, 33–44, 1984.

Litvin, F.L. and Zhang, Y., Robotic bevel gear train, *International Journal of Robotics Research*, **5**(2):75–81, 1986.

Litvin, F.L., Zhang, Y., Castelli, V.P., and Innocenti, C., Singularities, configurations and displacement functions for manipulators, *International Journal of Robotics Research*, **5**(2):66–74, 1986.

Luh, J.Y., Walker, M.W., and Paul, R.P., On-line computational scheme for mechanical manipulators, *ASME Journal of Dynamic Systems, Measurement, and Control*, **102**(2):69–76, 1980.

Ma, R. and Gupta, K.C., On the motion of oblique bevel geared robot wrists, *Journal of Robotic Systems*, **6**(5):509–520, 1989.

Ma, R. and Gupta, K.C., Signal flow graphs for spatial gear trains, *ASME Journal of Mechanical Design*, **116**:326–331, 1994.

McCarthy, J.M., *An Introduction to Theoretical Kinematics*, MIT Press, Cambridge, MA, 1990.

McMillan, S., Orin, D.E., and Sadayappan, P., Real-time robot dynamic simulation on a vector/parallel supercomputer, *Proceedings of the IEEE Conference on Robotics and Automation*, Sacramento, 1836–1841, 1991.

Meirovitch, L., *Methods of Analytical Dynamics*, McGraw–Hill, New York, 1970.

Mirman, C.R. and Gupta, K.C., Robot trajectory planning using the convolution operator, *Journal of Robotic Systems*, **4**(5):605–617, 1987.

Mirman, C.R. and Gupta, K.C., Identification of position-independent robot parameter errors using special Jacobian matrices, *International Journal of Robotics Research*, **12**(3):288–298, 1993.

Mooring, B.W., Roth, Z.S., and Driels, M., *Fundamentals of Manipulator Calibration*, Wiley, New York, 1991.

Oblak, D. and Kohli, D., Boundary surfaces, limit surfaces, crossable and noncrossable surfaces in workspace of mechanical manipulators, *ASME Journal of Mechanisms, Transmissions, and Automation in Design*, **110**:389–396, 1988.

Orin, D.E. and Schrader, W.W., Efficient computation of the Jacobian for robot manipulators, *International Journal of Robotics Research*, **3**(4):66–75, 1984.

Paden, B. and Sastry, S., Optimal kinematic design of 6R manipulators, *International Journal of Robotics Research*, **7**:43–61, 1988.

Paul, B. and Rosa, J., Kinematics simulation of serial manipulators, *International Journal of Robotics Research*, **5**(2):14–31, 1986.

Paul, R.P., *Robot Manipulators—Mathematics, Programming and Control*, MIT Press, Cambridge, MA, 1981.

Paul, R.P. and Stevenson, C.N., Kinematics of robot wrists, *International Journal of Robotics Research*, **2**:31–38, 1983.

Pieper, D.L. and Roth, B., The kinematics of manipulators under computer control, *Proceedings of the IFToMM Congress on Theory of Machines and Mechanisms*, Warsaw, **2**:159–168, 1969.

Pohl, E.D. and Lipkin, H., Real and extended workspace in robotic manipulators, *Proceedings of the ASME Mechanisms Conference*, Chicago, 1990.

Raghavan, M. and Roth, B., Inverse kinematics of the general 6R manipulator and related linkages, *Proceedings of the 1990 ASME Mechanisms Conference*, Chicago, DE**25**:59–65, 1990; also see *ASME Journal of Mechanical Design*, **115**(3), 1993.

Rastegar, J. and Deravi, P., Methods to determine workspace, its subspaces with different numbers of configurations and all the possible configurations of a manipulator, *Mechanism and Machine Theory*, **22**:343–350, 1987.

Rastegar, J. and Perel, D., Generation of manipulator workspace boundary geometry using Monte Carlo method and interactive computer graphics, *Proceedings of the ASME Design Technology Conference*, **3**:299–305, 1988.

Rivin, E.I., *Mechanical Design of Robots*, McGraw–Hill, New York, 1988.

Rosheim, M.E., *Robot Wrist Actuators*, Wiley, New York, 1989.

Roth, B., Performance evaluation of manipulators from a kinematic viewpoint, NBS #459, U.S. Government Printing Office, Washington, DC, 39–61, 1976.

Roth, B., Rastegar, J., and Scheinman, V., On the design of computer controlled manipulators, *On the Theory and Practice of*

Robots and Manipulators, Proceedings of the CISM/IFToMM Symposium, **1**:93–113, 1973.

Roth, Z.S., Mooring, B.W., and Ravani, B., An overview of robot calibration, *IEEE Transactions on Robotics and Automation*, **3**(5): 377–385, 1987.

Samak, S.M. and Gupta, K.C., Effect of dynamic model errors on robot precision, *Proceedings of the ASME Mechanisms Conference*, Chicago, DE**24**:73–77, 1990.

Silver, W.M., On the equivalence of the Lagrangian and Newton–Euler dynamics for manipulators, *International Journal of Robotics Research*, **1**(2):60–70, 1982.

Singh, V.K. and Gupta, K.C., A manipulator Jacobian based modified Newton–Raphson algorithm (JMNR) for robot inverse kinematics, *Proceedings of the ASME Design Automation Conference*, Montreal, **3**:327–332, 1989.

Stanisic, M.M. and Duta, O., Symmetrically actuated double point systems—the basis of singularity free robot wrists, *IEEE Transactions on Robotics and Automation*, **6**(5):562–569, 1990.

Stanisic, M.M., Pennock, G.R., and Krousgrill, C.M., Inverse velocity and acceleration solutions of serial robot arm-subassemblies using the canonical coordinate system, *International Journal of Robotics Research*, **7**(1):29–41, 1988.

Stepanenko, Y. and Sankar, T.S., A system approach to dynamic simulation of robotic manipulators, *ASME Computers in Mechanical Engineering* (CIME), **3**(6):61–68, 1985.

Stepanenko, Y. and Vukobratovic, M., Dynamics of articulated open chain active mechanisms, *Mathematical Biosciences*, **28**:137–170, 1976.

Sugimoto, K. and Duffy, J., Determination of extreme distances of a robot hand, in two parts, *ASME Journal of Mechanical Design*, **103**:631–636 and 776–783, 1981.

Taylor, R.H., Planning and execution of straight line manipulator trajectories, *IBM Journal of Research and Development*, **23** (4):253–264, 1979.

Tsai, L.W., The kinematics of robotic bevel-gear trains, *Proceedings of the IEEE Conference on Robotics and Automation*, Raleigh, 1811–1816, 1987.

Tsai, L.W. and Morgan, A., Solving the kinematics of the most general six and five degree of freedom manipulators by continuation methods, *ASME Journal of Mechanisms, Transmissions, and Automation in Design*, **107**:189–200, 1985.

Tsai, Y.T. and Orin, D.E., A strictly convergent real-time solution for inverse kinematics of robot manipulators, *Journal of Robotic Systems*, **4**:477–501, 1987.

Tsai, Y.C. and Soni, A.H., An algorithm for the workspace of a general N-R robot, *ASME Journal of Mechanisms, Transmissions, and Automation in Design*, **105**:52–57, 1983.

Tsai, Y.C. and Soni, A.H., The effect of link parameters on the working space of general 3R robot arms, *Mechanism and Machine Theory*, **19**:9–16, 1984.

Uicker, J.J., Denavit, J., and Hartenberg, R.S., An inverse method for the displacement analysis of spatial mechanisms, *ASME Journal of Applied Mechanics*, **31**:309–314, 1964.

Vijaykumar, R., Tsai, M.J., and Waldron, K.J., Geometric optimization of manipulator structures for working volume and dexterity, *Proceedings of the IEEE Conference on Robotics and Automation*, St. Louis, 228–236, 1985.

Waldron, K.J., Wang, S.H., and Bolin, S.J., A study of the Jacobian matrix of serial manipulators, *ASME Journal of Mechanisms, Transmissions, and Automation in Design*, **107**:230–237, 1985.

Walker, M.W. and Orin, D.E., Efficient dynamic computer simulation of robot mechanisms, *ASME Journal of Dynamic Systems, Measurements, and Control*, **104**:205–211, 1982.

Wampler, C.W., Manipulator inverse kinematic solution based on vector formulations and damped least-square methods, *IEEE Transactions on Systems, Man, and Cybernetics*, **9**(11):93–101, 1986.

Wampler, C. W. and Morgan, A.P., Solving the 6R inverse position problem using a generic-case solution methodology, *Mechanism and Machine Theory*, **26**(1):91–106, 1991.

Wang, L.T. and Ravani, B., Recursive computations of kinematic and dynamic equations for mechanical manipulators, *IEEE Transactions on Robotics and Automation*, **1**(3):124–131, 1985.

Whitney, D.E., The mathematics of coordinated control of prostheses and manipulators, *ASME Journal of Dynamic Systems, Measurements, and Control*, **94**:303–309, 1972.

Wolovich, W.A., *Robotics–Basic Analysis and Design*, Holt, Rinehart and Winston, New York, 1987.

Yang, D.C.H. and Lee, T.W., On the workspace of mechanical manipulators, *ASME Journal of Mechanisms, Transmissions, and Automation in Design*, **105**:70–77, 1983.

Yang, D.C.H. and Lee, T.W., Optimization of manipulator workspace, *Proceedings of the ASME WAM-Robotics Research and Advanced Applications*, 27–33, 1982.

Yashi, O.S. and Ozgoren, K., Minimal joint motion optimization of manipulators with extra degrees of freedom, *Mechanism and Machine Theory*, **19**(3):325–330, 1984.

Yih, T.C., On the C-B notation: An alternate homogeneous matrix method for the geometric modeling of lower pairs and its application to the kinematic modeling of spatial robots, *Proceedings of the ASME Mechanisms Conference*, Chicago, DE**25**:331–339, 1990.

Index

Mechanical Engineering Series *(continued)*

Laminar Viscous Flow
V.N. Constantinescu

Thermal Contact Conductance
C.V. Madhusudana

Transport Phenomena with Drops and Bubbles
S.S. Sadhal, P.S. Ayyaswamy, and J.N. Chung

Fundamentals of Robotic Mechanical Systems:
Theory, Methods, and Algorithms
J. Angeles

Electromagnetics and Calculations of Fields
J. Ida and J.P.A. Bastos

Mechanics and Control of Robots
K.C. Gupta

Wave Propagation in Structures:
Spectral Analysis Using Fast Fourier Transforms
J.F. Doyle